MINING TOWNS

ERIK EKLUND grew up in Wollongong and studied at the University of Sydney. He was appointed to the University of Newcastle in 1994, where, among other things, he was involved in a project to locate the convict coal mines that lay beneath the city. In 2001 Erik was a recipient of the Australian Academy of Humanities Travelling Fellowship to support the writing of *Steel Town: The making and breaking of Port Kembla*. That same year he held a visiting appointment at Georgetown University in Washington DC. In 2003 *Steel Town* was awarded the NSW Premier's History Prize. In 2005 Erik was a Visiting Fellow at the Humanities Research Centre, ANU. He joined Monash University in 2008.

MINING TOWNS

MAKING A LIVING, MAKING A LIFE

ERIK EKLUND

UNSW PRESS

A UNSW Press book

Published by
NewSouth Publishing
University of New South Wales Press Ltd
University of New South Wales
Sydney NSW 2052
AUSTRALIA
newsouthpublishing.com

© Erik Eklund 2012
First published 2012

10 9 8 7 6 5 4 3 2 1

This book is copyright. Apart from any fair dealing for the purpose of private study, research, criticism or review, as permitted under the Copyright Act, no part may be reproduced by any process without written permission. Inquiries should be addressed to the publisher.

National Library of Australia Cataloguing-in-Publication entry
　Author: Eklund, Erik Carl.
　Title: Mining towns: making a living, making a life/Erik Eklund.
　ISBN: 978 174223 352 9 (pbk)
　ISBN: 978 174224 111 1 (epub)
　ISBN: 978 174224 368 9 (Kindle)
　ISBN: 978 174224 598 0 (pdf)
　Notes: Includes bibliographical references and index.
　Subjects: Mineral industries – Australia – Social aspects.
　Miners – Dwellings – Australia.
　Cities and towns – Australia.
　Dewey Number: 307.7660994

Design Josephine Pajor-Markus
Cover 'Miners finishing a shift at Western Mining Corporation's Silver Lake shaft at Kambalda in south-east Western Australia, 28 March 1969'. NLA. It has not been possible to trace the names of these men.

The author plans to donate royalties from this book to the Royal Flying Doctor Service, which is active in all the regions covered in this book. See <www.flyingdoctor.org.au>.

Contents

Introduction 1

1 The global and national context 9

2 Broken Hill: Icon of working-class culture 35

3 Mount Morgan: In the thrall of modernity 69

4 Queenstown: 'They've got to come here and they've got to learn about it' 105

5 Port Pirie: 'Essentially hard and practical' 135

6 Mount Isa: Normalising outback suburbia 173

7 Kambalda: Modernity, environment and experience 205

Postscript 233

Appendix: The oral history component 240

Notes 243

Bibliography 273

Acknowledgments 289

Index 291

Map of Australia showing the six towns covered in this book and the capital cities

Introduction

Imagine a map of Australia with the capital cities marked in bold and the regional cities in smaller print. This kind of map reflects a hierarchy which many histories have followed. Political, social and economic histories focus on the centres of power and commerce – Melbourne, Sydney and Canberra. Beliefs about value and importance shape all too clearly the 'where' and the 'who' of the Australian historical record. Moving beyond a city-centric view of history, and changing the starting point for analysis, reveals different aspects of the Australian experience. Rather than heading to the familiar east coast with its ever-expanding conurbations, we will look elsewhere. This book will turn that hierarchy on its head, and we will find that 'elsewhere' is actually somewhere: the places, communities, and lives beyond the 'Big Smoke'.

Elizabeth Firth was born in Broken Hill in 1933. She is a busy, socially engaged woman who is proud of her town and of her mother, a pioneering feminist in the Broken Hill branch of the Labor Party. Her father was born in Broken Hill of Victorian parents. Her mother came to Broken Hill from the copper mining town of Kadina in South Australia. Her father worked underground at the Zinc Corporation as a timberman, and before marriage her mother worked at a local retailer, J.C. Goodharts. Asked about her earliest memories, she recalls an uncle who was injured at the mines:

> He had a motor car and he used to take me all over the place. I was his little girl because he didn't have any children. And I remember going to the pay office at the Zinc Corporation every fortnight with him to pick up his workers compensation pay. And there was tuppence in the pay envelope and the tuppence was mine ...

Mining Towns

> I think I was about three when I remember that and I remember a few outings we used to have.[1]

In 1975, on the other side of the continent, Will Manos and his family arrived in Kambalda, on the goldfields south of Kalgoorlie. His family had moved there from Darwin, via Perth. His grandparents were living in Kambalda East. Will was five when they arrived: 'Thirty-five years later my family are still in Kambalda and we love the place to death and we certainly call Kambalda home.'[2]

From the old mining town of Broken Hill in far western New South Wales to the nickel mining town of Kambalda in Western Australia, these two stories are a tiny selection of many individual stories from beyond the 'Big Smoke'.

The inspiration for this book comes from over 30 detailed interviews, and numerous other informal discussions, with people from industrial and mining towns in regional Australia. These interviews, varied though they are, have common themes: migration and movement, and place-specific identity. The book reveals evocative memories of place, complemented and counterbalanced by stories of movement, migration and connection with other places.[3] These other places could be mining towns, ports, or places of origin for migrants. The capital cities are relevant too, because while they are not the focus of this work, they play an important role as sites of financial and political power. Beyond them, communities have developed deep links to their locality, but these towns are also connected to wider regional, national and international patterns.

This book examines the history of six Australian towns which rarely figure in national histories or national discourse: Broken Hill (New South Wales), Mount Morgan (Queensland), Queenstown (Tasmania), Port Pirie (South Australia), Mount Isa (Queensland) and Kambalda (Western Australia). I hope it encourages us to think anew about these places, and to question a city-centric national history. It explores

Introduction

another way of framing our past – as a network of place-based histories and experiences criss-crossing the continent. The book is both a detailed engagement with the specific histories of six towns and a contribution to a more geographically inclusive national history.

Three of the towns were a product of the 1880s and all were metal mining sites. Broken Hill in far western New South Wales mined silver, lead and zinc from 1883 and became one of the richest mining fields in the world, the progenitor of major Australian companies. Queenstown grew in the gold, copper and iron mining fields in and around Mount Lyell, near the west coast of Tasmania. Gold mining began in this area in 1881 and the town of Queenstown rose from the ashes of a major fire which burnt out a prior settlement, Penghana, in 1896. Mount Morgan was a gold (and later gold and copper) mining town in Central Queensland which began production in 1883. By the 1890s, all three of these mines were operations on a scale that was unprecedented in Australia.

In contrast to the inland metal mining towns of the 1880s and 1890s, Port Pirie is near the northern end of South Australia's Spencer Gulf, and is not, in fact, a mining town. A trading port which first appeared in the 1870s, by the 1890s Port Pirie had an important economic link with Broken Hill, with the establishment of smelters treating Broken Hill ore. Port Pirie reveals important connections between the mining and industrial sectors, as the products and profits of mining reached out to affect other places and industries.

Finally, there are two 20th century towns. Mount Isa, in north-western Queensland, began as a silver and lead mine in 1923, and later developed a large smelting and copper mining operation. The town grew around the mine site and by 1930 on-site industrialisation, including a large lead smelter, was in place. Kambalda, on the eastern goldfields of Western Australia, had a short life as a gold mining town from the end of the 19th century. It reappeared in 1966, becoming a significant nickel producer and the site of a major modern company-designed town.

These towns cover every state in Australia except Victoria, and range from mining to industrial functions. Some are isolated inland localities, others are closer to coastal areas. Some developed through the joint efforts of companies, governments and municipal authorities, others were company towns for important periods of their history. Coverage is very selective In fact the more one considers the omissions, the more weight there is to an argument that a synthesis of the history of these towns is long overdue, and is something beyond the scope of what a single scholar can do in one book.

That the experience of regional Australians is not well represented in our national history indicates something about the way history is conceived in Australia. This highlights the second reason for writing this book: to flesh out our national story. General histories of Australia are often focused on the experiences and events of the capital cities. The assumption seems to be that the lives of city-dwelling urbanites are what our history is about. The more isolated regional areas are thought to have taken their cues from the big cities, with historical change happening in ways that saw patterns and experiences flow out from the city to be replicated in smaller and less dramatic ways in the regions.

More recent work by historians, geographers and others tells a much more interesting story. Historical change occurs in fragmented ways, both chronologically and geographically. In each place the forces of history come together in unique ways: in varying regional contexts; through a particular group of settlers and Indigenous people; via different state policies; and in distinctive company traditions and approaches. Historians J.W. McCarty, and more recently Lionel Frost, have proposed the concept of 'inland corridors' as useful in this kind of history, and Alan Mayne and others have written of history 'beyond the black stump' or 'outside country'.[4] These new ways of conceptualising the geography of Australian history acknowledge the peculiarities of place but also chart connections and contexts, bringing place and history together in compelling ways.

The third reason for writing this book goes to the heart of why we research and write history. A study of the past provides a breadth and depth of understanding that can inform the present. A society which

Introduction

loses its historical compass is a society with a short collective memory, cut loose from tradition, vulnerable to abuse and exploitation, heedless of the legacies of the past.

As present-day society struggles with the shape and consequences of a major mining boom which began in earnest in the early 21st century, it is timely to look back and consider how past generations have fared. The focus of this book is not so much economic and financial issues, which dominate our current debates, but the process of community formation. How did people create societies in isolated locations with few resources? How did 'ordinary people' make sense of their worlds? How did they respond to rapid economic change, find identity and attachment in these remote areas? What were the patterns of inclusion and exclusion? What were relations between Indigenous and non-Indigenous people like? This last question is a crucial one because many of these towns have important Indigenous histories, and although I have highlighted these issues where research material was available, more detailed research is required.

The answers to these questions are all valuable lessons we can draw on. Not all of the lessons will inspire great pride. In fact the lessons from these towns often reveal narrow thinking, and a privileging of profit at any cost. But in among the painful and often confronting patterns of racism, industrial conflict and exclusion, we also see loyalty, attachment and affinity, including times and places where societies mobilised effectively, and where people felt a sense of belonging and led prosperous, fulfilling lives.

Searching through the minutiae of history at a community level helps us understand how these societies coalesced, what mattered most to people and what gave them a sense of satisfaction. In my research I found that great activity and energy was directed towards the civic and voluntary sectors. Here many residents found solace from their hard working lives. They sought change and improvement in their towns, maintained and expressed culture from other places, and just enjoyed the company of others in a round of dinners, dances, picnics and socials.

Yet many of the communities I visited were under siege from the present-day mining boom. Working arrangements – principally a fly-in fly-out or drive-in drive-out workforce, on 12-hour shifts – were sapping the life out of them, and undermining the voluntary and civic sectors. If anything, we appear to be more utilitarian, and even more focused on making money than those who rushed to the boom towns in the 1880s. When these communities began life, they were difficult places to live. Working conditions were dangerous, and their social structures were often rudimentary, but as time passed they became more civilised and their worst features were moderated. By as early as 1920 underground workers at Broken Hill had achieved a 35-hour week, for example, and the town had a full and vigorous social and political life as a consequence. The reversion to 12-hour shifts in the mining industry from 1991 made many of the union men I interviewed shake their heads in disgust.

This emptying out of social life has real consequences. We need to decide whether we are comfortable with communities made up of highly mobile but ultimately dislocated people who have no time or energy left for their civic, social and political lives.

Our 21st century mining boom will not create towns in the desert, just airstrips and temporary camps. Imagine an Australian continent without the rich legacy of Broken Hill, or the distinctiveness of Queenstown. This is what we are hurtling towards. So it is time to pause and reflect on the past, on how communities formed and developed, survived or disappeared through the multiple booms and changes that have affected the Australian economy since the great regional boom precipitated by the gold rushes.

The sheer complexity and diversity of the past becomes history through the intervention of the historian. Our own perspective shapes what we see and how we see it. I grew up in Wollongong, where the BHP steelworks and the coal mines dominated the regional economy. These industries shaped the demographics and cultural makeup of

Introduction

the Illawarra, which had its very own north/south divide. While large numbers of European migrants settled south of Wollongong and found employment at Port Kembla, the coal mines to the north attracted British or locally born workers.

After studying the history of the Illawarra, and of the steel town of Port Kembla in particular, I became more interested in the connections between industrial and mining towns. Stories shared about Port Kembla often featured other towns. Newcastle came up, as did the Hunter Valley coalfields, Broken Hill and even Mount Morgan. In 1999 I started teaching a history subject on Australian mining and industrial towns, and the responses of mature age students in particular, some of whom had had first-hand experience of the boom happening in regional Australia, confirmed my belief that there was another Australia out there, with older patterns of movement and mobility. The path of working for five years in an isolated mining town and then returning to the east coast was a well-established one, as familiar to the South Australian and Victorian miners who moved to far western New South Wales in the 1870s and 1880s as it is now, though in vastly truncated form, to today's fly-in-fly-out workers.

While on leave in 2004 I met grey nomads who had not only travelled all over Australia but had also worked in multiple places, from Mount Isa in the northwest of Queensland to the goldfields of Western Australia. Their lives were a long way from the halls of academia, and from the café culture of capital cities. I found this experience both exciting and unsettling. I began reading research about these places, and I noted the absences, the way national discourse allowed these towns to be sites of economic growth or industrial conflict but little else. What of their particular social characteristics? What of their own regional histories? What kind of history could acknowledge the experiences of the residents of these places?

To answer this question I travelled to these places, sampled life there and interviewed locals, translating the regional sense of place I found on all my trips into this book. Ultimately it is my interpretation in the pages that follow, but I have tried to draw on the voices and memories of local people. I wanted to find a way to acknowledge their histories. I wanted to see what Australian history looked like if you lived in Mount Morgan or Mount Isa, and to see the world as my informants saw it. In this book we figuratively dive into the Australian past through these places.

Before we begin this town by town journey, I have provided some overall context, in Chapter 1. The other chapters tell what I found on my long journeys to places that are seen by many as the fringes of the Australian experience, but for whose residents are the centre of their own rich and compelling lives. On some of these journeys I was alone, and on others I travelled with my own family, in a faint echo of those past journeys that saw thousands travel to new places, establish new lives, and build communities beyond the 'Big Smoke'.

1
The global and national context

This chapter considers the political economy underpinning the development of the six towns the book examines. It reveals the connections that flow through these localities to regional centres, to state capitals and beyond: financial and investment connections, flows of union and community organisation and of people. Across the continent, mining towns have become nodes for capital, labour and culture. In turn, the towns and their dominant companies have influenced other places, such as port communities and nearby regional centres. The picture is of an interlacing of geography, finance and history.

Mining and industrialisation

Colonial economies developed as almost separate hubs around foundational towns. From Sydney the pastoral economy moved west of the Blue Mountains, branching out inland, north and south, in the 1820s. The coastal areas were another area of growth, with towns such as Newcastle and Wollongong growing from the 1820s and 1830s. The scenario was similar in Van Diemen's Land, where the sites of first settlement – Hobart and Launceston – became economic hubs which captured and controlled inland development. Growth was predicated on the dual expectations of a convict or ex-convict workforce and the expropriation of the resources of the land and sea from Indigenous peoples.

Mining Towns

There were patches of mining development that had particular effects in regional Australia. From 1803, Newcastle, in New South Wales, developed as a scattered collection of pit-top coalmining villages around the transport hub of the harbour. Copper mines in South Australia first developed in the 1840s at Burra and Kapunda, north of Adelaide; the Glen Osmond silver-lead mines, closer to Adelaide, developed in the same decade.

Gold changed everything. Images of these years, in the second half of the 19th century, evoke a frontier-like, anarchic period of development. Flows of free migrants meant a significant increase in the size of the domestic market – the Victorian population grew from 97,489 in 1851 to 538,628 in 1861, for instance – as well as changes to the labour market, society generally, and politics. By 1860 convict Australia was a receding memory in most eastern colonies. Gold mining and urban manufacturing (mostly in the colonial capitals) marked the start of the move of Australia's economy towards more technologically adept and capital-intensive industries.

Many mine sites developed on-site processing, and all required harbour or rail infrastructure as well as people with engineering skills to make these complex mechanical systems (which increasingly replaced manual labour) function effectively. The final major expression of gold fever occurred in Western Australia – a colony that had struggled since its establishment in 1829 – from the late 1880s. It was the last gasp of the lone digger, and after the initial rushes, larger companies, with their more technically sophisticated operations, began to dominate in Western Australia much as they had done in the east.

As well as gold, there were other mineral discoveries across the continent. In the 1860s, major finds occurred in the 'copper triangle' on South Australia's Yorke Peninsula: Wallaroo, Moonta and Kadina. Tin mining developed rapidly in the 1870s in the New England region of New South Wales, in northern Queensland and the Darling Downs, and in the northwest of Tasmania. Australia was the world's largest tin producer in that decade.

By the 1870s, clearer patterns of mining development had emerged,

The global and national context

and it was clear that Australia was rich in mineral resources. There was gold, but there were also coal, tin, silver and copper. Geology was shaping the continent's patterns of urban settlement. Geology also helped create distinct regions. It was rare for an ore body not to have payable lode right across a geological 'province'. The Mount Isa ore body, for example, is estimated to cover 50,000 square kilometres. This accounts for the number and diversity of mines in the region, and influenced regional and town development from the earliest copper mines in the 1860s through to more recent development at the Cannington and Century mines.[1]

It was also apparent by the 1870s that larger companies, with their more sophisticated mining methods, were taking over. They attracted significant investment from Britain. After 1872, the newly operational telegraph made regular communication with British-based boards and investors more feasible. These international links also flowed through the colonial capitals, which controlled the mining economy both financially and politically. Thus the scene was set for major expansion of the mining industry in the 1880s, almost invariably in isolated areas of regional Australia, but still bound to the city.[2]

Such rapid and large economic expansion needed a labour force. The 'long boom' between 1860 and the late 1880s, with its steady annual economic growth and active colonial governments, created a new image of the Australian colonies – as places of opportunity, rather than open prisons. After 1851 the widespread clamour for gold, and its promise of a new life and new opportunities, led to increased migrant flows to the colonies. Waves of arrivals, some with mining backgrounds, were coming, forming the core of a mining workforce that would travel from field to field in search of employment.

There were two countervailing forces operating on mining labour markets. New labour-saving technology reduced labour demand at mining sites and smelting operations. At the same time, however, major expansion and new projects, along with employment in related industries, generated new demands for labour, especially in the latter half of the 19th century.

The industry was, however, highly erratic and unpredictable. The excitement engendered by opening a new mine, which would provide employment for hundreds of men, was tempered by the real possibility that operations could close at any time. Any one of a number of problems could hamper or halt production: technical problems, a shortage of capital or equipment, accidents, bad weather, or a drop in metal prices. So while labour demand for new projects was steadily growing, the location of the demand tended to be ever-changing. This created a group of itinerant workers who moved from field to field, mirroring the itinerant workers of the pastoral industry. The few major mining fields with greater longevity, such as Broken Hill and Mount Lyell, were the exceptions. Yet even here, downturns could send workers to other mining centres or city labour markets for extended periods.

By 1900, mining contributed 20.1 per cent of Australian Gross Domestic Product, and was almost as valuable (at £23.8 million) as the rapidly growing Australian manufacturing sector.[3] It had contributed a mere 6.8 per cent in 1880. Three of the towns under focus here – Broken Hill, Queenstown and Mount Morgan – had their genesis in the 1880s.[4]

There were two factors that facilitated this late 19th century Australian mining boom. First, the lessons learned on the goldfields were often a basis for later mining developments. The School of Mines at Ballarat and Bendigo, and later Kalgoorlie, trained highly effective mining engineers. Locally trained men were supplemented by a steady stream of British and German specialists. From the 1890s, US experts were also common, as Australian managers began to recognise the skills and experience available from the large and advanced North American mining sector. Federation, too, was a stimulus to industrial development, as it created one common market across the Australian colonies; this helped recovery from the depression which had begun a decade earlier.[5]

Second, colonial governments were strong supporters of mining development: the sector was lightly regulated but greatly helped by government investment in transport infrastructure. Reform in company legislation, pioneered in Victoria from the 1850s, also helped, as it gave potential investors some protection.[6] The 1890s depression, and the consequent widespread loss of investor confidence, led to another round of legislative innovation. Many colonial politicians doubled as investors and were not averse to serving on the boards of companies that received government assistance or support.

However, inconsistency in both policy and delivery meant differential access to infrastructure and state support. Many regional towns struggled against the strong centralising force of state governments. In New South Wales, differential rail freight rates favoured the port of Sydney over Newcastle and other regional ports. In Western Australia, the state government quickly built rail lines from Perth or Fremantle to the goldfields, but it was slow to build lines that linked the goldfields to closer ports. Despite many years of lobbying by port town Esperance and goldfields representatives, the line from the gold town of Norseman to Esperance was only completed in 1927, and a new standard gauge freight line to Kambalda was completed in 1974.[7] At the same time significant amounts of public money poured into capital city infrastructure, with roads, bridges and railway lines connecting the city with its hinterland and beyond. This made it very difficult for regional towns, especially ports, to compete as transport hubs with capital city locations. In Queensland, though a more decentralised political economy modified this picture, city-focused development persisted well into the 20th century. These frustrations drove the formation of a 'country party' in all states.

Economic historians Barrie Dyster and David Meredith posit two distinct economies in Australia – rural and urban – with few links between them.[8] Yet these towns had strong region-to-city connections. For a start, city-based investors were the dominant group of shareholders in mining companies, and city-based directors and financiers in Melbourne, Sydney, Brisbane and London controlled them. Those shareholders and directors were supported by city firms of solicitors, secretarial services, insurance agents, shipping companies, trading houses, accountants and mining consultants.

Moreover, these connections were not merely between the city and the bush. The development of Port Pirie and Queenstown reveal connections across the regions, as mining capital and technology flowed from one place to another. As the most convenient seaport, Port Pirie, like many other ports, benefited substantially as goods travelled along the trade route following the rail line into far western New South Wales. Queenstown was an offshoot of BHP's success at Broken Hill, and it, like Broken Hill, was controlled from Melbourne. The development of Queenstown assisted Strahan, which was the closest port for both Queenstown and the mining town of Zeehan.

For Mount Morgan the relationship to nearby Rockhampton was crucial. Kambalda and Kalgoorlie were similarly mutually dependent, and both had financial and management links to Melbourne and Perth. Thus it can be argued that the economic and financial geography of the economy is more closely aligned with Frost's model of corridors linked by trade and transport, than with Dyster and Meredith's model.[9] In the 1890s, if the colonial capitals were the economic hubs, the mining towns were important spokes. They were a focus of considerable economic activity in an otherwise depressed financial environment.

By the 20th century, the nature of mining industry growth changed. New projects were less transitory; they were more carefully crafted and planned by what were by now larger companies. The inexorable reduction in labour demand began to have an effect. The high-water mark for mining employment was 1911, with 87,000 employed.[10] The Great War gave a further stimulus to Anglo-Australian capital as German financiers such as Aaron Hirsch und Sohn, who had been heavily involved at Mount Morgan and with the Port Kembla copper refinery, were ousted from the Australian base metals industry once war began. This was the work of Collins House financier W.S. Robinson, in concert with the then Federal Attorney-General, William Hughes.[11] Protected by imperial preference, Australia found a ready market in Britain for not only wool and wheat, but also base metals such as lead, copper and zinc. The turning point was 1915. That year Broken Hill Associated Smelters was formed as a major component of the Collins

The global and national context

House group, and Australian steel making began on a large scale in Newcastle. Like the mining operations at Queenstown, the Newcastle works was built on Broken Hill profits.

After the Great War, British companies and investors increasingly retreated behind the walls of the Empire. As historian Peter Cochrane notes, '[t]he Empire became a haven for a battered old power'.[12] A consequence of this was the relocation of entire sections of industry to Australian bases. John Lysaght's, the Bristol-based English firm, moved to Newcastle in 1921. In the 1880s state governments had built railways; in the 1920s, they built infrastructure for electrification. During that decade, new generating facilities, electricity supply, tram and inner-city train systems were constructed. This was a great boon not only for mining, but also for local copper and steel manufacturers, and for workers in metal fabrication plants, and builders of electrical generators. Many materials were sourced locally, and in some cases, state government contracts actually stipulated the use of Australian-made materials.

From 87,000 in 1911, the Australia-wide mining workforce declined to 59,000 by 1933, and to 37,000 by 1981.[13] At large operations such as Broken Hill, there were 4452 miners by 1933, compared with 11,000 in 1892. Often these reductions were achieved with a concurrent increase in production figures – the dream of every capitalist manager and shareholder. Mechanisation, open cut mining methods, and new machinery – aerial ropeways, conveyor belts, diesel trucks, mechanical diggers and the like – all contributed to the labour reduction. Efficiencies also extended to innovation in treatment methods, which was particularly crucial in the smelting industry. From 1890 to 1915, a host of new metallurgical processes were trialled. Some succeeded spectacularly; others failed. Metallurgy was an inexact science and new ore types all behaved differently in the smelting and refining process. Complex ores could contain commercial quantities of lead, zinc, gold and silver, as well as traces of other useful metals. The most famous metallurgical challenge was solved by the Zinc Corporation, through the invention, and commercial proving, of the flotation process.¹⁴ From 1905, this allowed the large-scale reprocessing of otherwise wasted Broken Hill tailings that were rich in zinc, and so spawned another major player along the line of lode.

It might be reasonable to expect that colonial economies of the hub and spoke model were significantly altered by these wide-ranging 20th century developments. But high transport costs, generally inadequate rail infrastructure (including change of gauge problems), and a small road transport industry meant that the capital city/regional hinterland relationships of the 19th century were not significantly changed until the 1960s. Established patterns, with existing trade routes and transport corridors, were remarkably durable.¹⁵

What did change post-1945 was geological surveying and drilling. Much of the progress in these areas had been pioneered by Collins House and BHP. But after 1945 BHP was no longer a player at Broken Hill or at Port Pirie, and the Collins House empire began to re-form as some of the dominant players retired or died. W.L. Baillieu had died in 1936 and Colin Fraser in 1944. In addition, key Collins House companies merged with British companies, and through various changes in name and ownership became the Conzinc Rio Tinto Group, later CRA and then Rio Tinto. US money also slowly replaced British capital. The Empire became the Commonwealth and lost much of its trading and financial coherence as Australia's major trading partners changed, first to include the US, then Japan, and more recently China and India.

Geography and the economy

The picture of the changing economy sketched above reveals crucial connections to the six towns examined in this book. All too often we are encouraged to think of the economy as an abstract structuring factor in our lives. What is clear, however, from an account of mining and industrialisation in Australia is the geographical basis of the economy: it is made up of real places and relationships between them. From the 1880s, when the first towns considered here developed, large-scale financial investment in Australian mines was common, and mining was increasingly capital intensive. Whereas a few diggers could work an alluvial claim with little or no start-up capital in the 1850s, by the 1880s raising money through share issues was typical. In the 1870s and 1880s, regional capital played an important role, particularly in the start-up phase, but by the late 1880s, the independent digger who worked a claim on his own was the subject of nostalgic reminiscence by many in the labour movement. Wage labour had become the norm. The last gasp of this popular dream of individual independence was the West Australian goldrushes, especially at small finds such as Kambalda, where lone prospectors once again led the way. However, there they were replaced by a large company within a year. The still commonly expressed preference of 'working for myself' is a modern-day echo of this desire for liberation from wage labour. The emotional and political burden of adjustment to a life of wage labour lies at the heart of 20th-century experience for workers in these towns, and perhaps in all others too.

Broken Hill

In 1883, a boundary rider and amateur geologist, Charles Rasp, pegged a claim on a ridge in 'the Broken Hill paddock', part of the Mount Gipps pastoral run west of the Darling River. Rasp, a German immigrant with a shady past and at least two other identities, was employed by the Mount Gipps station, where the manager, George McCulloch, along with others, formed a syndicate to develop the mine. The syndicate formed BHP in August 1885. Key players included men such as Bowes Kelly, who was present at the formation of the company and served as chairman of the company in the 1890s and again after the Great War. Kelly had worked as a station manager, and dabbled in land speculation. William Knox, an accountant and experienced rural businessman, still had excellent Melbourne connections, and as the company's first secretary was 'regarded as the brains behind BHP'. All these men shared the lucky combination of perfect timing and shrewd investment. With capital behind them, they could maintain their investment in mine development until the profits flowed. For Kelly, the original £150 share purchase in 1884 was worth £1.5 million in ten years. He, along with the other early investors, joined the ranks of the 'Broken Hillionnaires'. These regional capitalists soon relocated to Melbourne, where they already had family and business connections. In 1901 Kelly purchased 'Moorakyne', a large property in East Malvern, Melbourne, originally designed by architect Charles D'Ebro. This was Kelly's 'region', and he was as engaged in it as the miners at Broken Hill would be in theirs. Kelly was a councillor of the East Malvern Shire from 1892 to 1896. Sweeping views of the distant ranges framed the everyday scenery of this well-to-do suburb; the wealth that underpinned his lifestyle was mined a world away.

Other early investors moved to London to live the high life on the line of lode's fabulous profits.[16] George McCulloch, as noted, had been manager of Mount Gipps station, north of Broken Hill, beyond Stephens Creek. His uncle, James McCulloch, a Victorian politician and business entrepreneur, was its owner. After BHP was floated in 1885, George McCulloch became a director on the board. The Broken Hill find transformed his life. William Corbould, who was a young assayer in Broken Hill in 1886 and 1887, recalled that McCulloch moved:

> to the mansion in Queen's Gate, Kensington, London, where I met him again years later. The Proprietary had made him a millionaire and he lived amid his collection of art, attended by a retinue of servants.[17]

McCulloch's art collection was famous, his penchant for English and French artists renowned.

Regional players like Rasp and McCulloch were less important after the heady days of the 1880s. The wealth of the Broken Hill line of lode was controlled largely from Melbourne and London. The Broken Hill find proved so rich that it quickly drew in capital. The bonanza lode discovered in 1886, a rich vein of silver-bearing ore, helped secure initial capital and maintain hopes until more investors had been brought on board. In the end, it was largely Australian, not British, capital that arrived. There was no regional centre near Broken Hill that could form a conduit between the town and the potential investment community, so this role was performed by Melbourne and, to a lesser extent, Adelaide. More importantly, by the 1880s and 1890s large amounts of capital investment, along with more formalised and corporatised procedures for registering companies on sharemarkets, were required. The days of regional capital financing regional industrial and mining development were passing, and the increasing centralisation placed greater control in the hands of city-based financiers.

Broken Hill contained a substantial silver-lead-zinc lode that bankrolled the increasingly dominant BHP. The town of Broken Hill developed as a multi-company mining field, unlike the predominantly single-company fields of Mount Morgan, Mount Lyell and Mount Isa. BHP was the first, and the most important, company in the town until its eventual withdrawal in 1940. However, long before then BHP had sold off the northern and southern ends of its mining lease – to companies that themselves became major mining players in Australia: Zinc Corporation, Broken Hill South and North Broken Hill, among others. By 1915, these non-BHP companies coalesced around an interlocking group of companies – known as the Collins House Group – with common shareholders and directors. Together BHP and the Collins House Group dominated the Australian mining industry for 50 years.[18]

Mount Morgan

The move from regional syndicate to major national company that occurred in Broken Hill was mirrored at Mount Morgan, in Queensland. Local commercial interests in nearby Rockhampton were fundamental to founding the mine and to providing its initial capital. The Morgan brothers, local graziers who had taken over the selection where much of the orebody was located, invited Rockhampton interests to join the syndicate, which was formed in 1882. The syndicate included Thomas Skarrat Hall, a bank manager from Rockhampton,

William Knox D'arcy, a Rockhampton-based solicitor, and William Pattison, a local grazier. Each contributed £2000. The syndicate continued, in various forms, until 1886, when the company was formed. While the mine soon established links to Sydney and Melbourne, its Rockhampton capital base was a distinguishing feature of its first 20 years of existence. The ability of the Rockhampton business elite to become fully fledged investors in a major mining venture was partly due to the town's core of wealth, itself generated by pastoralism, plus the meat processing and transport industries that developed around it in the 1860s and 1870s. The initial flush of gold-related development in the 1850s and 1860s played a role too.

Regional capital was the basis of many mining concerns founded in the 1870s and 1880s. The iron and steel industry at Lithgow (New South Wales) was supported by regional capital, as were copper smelting at Cobar (New South Wales) and large-scale gold mining at Charters Towers (Queensland) and Ballarat (Victoria). In the 1870s, profits from Newcastle's Lambton coal mine financed the construction of two copper smelters at the Adolphous William mine in central Queensland.[19] In the late 1800s these regional capital flows criss-crossed the continent. And it was through these regional elites that the international and capital city investment flows met these new mineral finds.

Mount Lyell/Queenstown

On the isolated west coast of Tasmania there were few investors. Even the pastoralists and station managers, who were often in the right place at the right time in other locations in the 1870s and 1880s, were absent in this region, as there were few rural industries. Gold, tin, silver, zinc and lead mining had developed on the west coast from the 1870s, and some wealth flowed through to the burgeoning town of Launceston. A gold find in 1883 around what was to become Queenstown first raised interest in the area. Capital was sourced principally from Broken Hill. Directors and senior managers from BHP such as William Knox, William Orr, Bowes Kelly and Hermann Schlapp, entered into the Mount

The global and national context

Lyell field using their financial connections, technical expertise and, above all, healthy profits from Broken Hill. The Mount Lyell company was floated in 1893, in what was a difficult climate for attracting investors: the target was not regional capital from Launceston, but investors from London. Bowes Kelly and William Knox visited London in 1893 and assiduously worked the financial and political networks. International capital markets were against them, however, and a severe depression, coupled with a decline in base metal prices, made it impossible to attract investment to an obscure Tasmanian mining venture. Instead, a loose collection of Melbourne-based investors and directors financed the new company. Without domestic capital formation these mines would have failed.

The origin of investment capital was crucial for these towns. As rural economists and sociologists looking at contemporary Australia note, small towns are unable to maintain capital flows in their areas. Money, trade and profits are generally siphoned off through city-based financial institutions, and larger companies. Where regional capital is mobilised, as it was in Rockhampton, greater wealth flows through the whole region. It also makes locals feel more ownership of, and commitment to, local development. And local boards of directors have greater regional knowledge. Regional capitalists can allow in some outside capital, but often, because of their own regional interests, they make decisions that allow greater regional capital retention.

In a later period, when major industrial assets were established near relatively prosperous commercial centres – Port Kembla's iron and steel industry, only a few kilometres south of Wollongong, for example – development proceeded along substantially different lines. The nature of capital formation and company structure had changed so much by the 1920s that regional capital formation was largely impossible. Wollongong business leaders were completely marginalised when Australian Iron and Steel was formed in 1928, by national and international firms. This had ramifications for more than personal wealth; it also had impacts upon the community and social wealth. It meant that the population's relationship to local industry was as consumers and workers. There was no opportunity for a more diverse and complex relationship of capital to place. It also brought a sense of powerlessness, as corporations acted at geographical scales much larger than any local society and economy. The ways of the 1880s and 1890s had changed substantially by the 1920s and 1930s, when Mount Isa Mines and Western Mining Corporation were formed.

Port Pirie

As a port town and an important transport hub, Port Pirie's development was different from that of the mining towns, and it exemplifies the follow-on effects of major mineral development. The town developed rapidly from the 1870s as a wool and (later) wheat trading port.[20] Underpinned by an expanding agricultural hinterland, and a potentially valuable harbour, the town's political and social elite ensured its prosperity by securing important railway links with inland wheat-growing areas in the 1870s and 1880s, and with Broken Hill by 1888.[21]

While one option for mining companies was on-site smelters and processing plants, and a number of mining companies developed these, the other option was to build processing facilities at or near port locations with proximity to other required inputs, such as coal to fire the hungry furnaces. In 1895, the Sulphide Corporation, sourcing its silver-lead-zinc ore from the Central Mine at Broken Hill, set up smelting operations at Lake Macquarie, south of Newcastle. In 1907, the Mount Morgan company, along with other Queensland mining firms, underwrote the formation of the Electrolytic Refining and Smelting Company at Port Kembla. In both cases, there was abundant coal and good rail and harbour facilities which made it viable to ship the ore so many kilometres for processing.

For Broken Hill companies, moving away from on-site smelting had other advantages. From the 1890s, the smelting operations located on the Broken Hill line of lode showed alarming signs of 'creep', or ground movements, which led to cracking in their foundations. In addition, moving away from Broken Hill gave management the opportunity to avoid dealing with the effective and occasionally militant union organisations in Broken Hill. Once the rail line from Port Pirie to Broken Hill was complete, the arguments for a Port Pirie base were even more compelling. This rail corridor carried supplies – timber props and equipment – across to Broken Hill, and large tonnages of ore back to Port Pirie.

The British Blocks Company began operating its furnaces at Port Pirie in 1889. BHP initially leased furnaces from this company, but from 1892 they began establishing their own large works on-site. By 1896 the two smelting companies, and various coke-making facilities, employed approximately 1000 men at Port Pirie.[22] By this time BHP was convinced of the advantages of smelting at a seaboard location, and decided to shift its entire smelting operations from Broken Hill to Port Pirie. In 1904, BHP smelters at the port town employed an average of 1070 workers for the year, and by 1913 this figure had increased to 1500.[23] By 1915, the Port Pirie smelter was the largest in the British Empire.

In that year, four major Broken Hill-based companies – BHP, Broken Hill South Ltd, North Broken Hill Ltd and the Zinc Corporation – established the Broken Hill Associated Smelters Pty Ltd (BHAS). BHAS was a co-operative venture organised to fill the gap left by the wartime displacement of German firms from the Australian base-metals industry.[24] These Broken Hill companies (except BHP) had similar labour management philosophies, and had senior directors in common. BHP was rather marginal in the group, and sold its holding to the Zinc Corporation in 1925. Significantly, BHAS, part of the emerging Collins House Group, became a focal point for the analysis of and dissemination in Australia of new labour management strategies. BHAS, a major lead supplier to the British muniitions industry during the Great War, was exposed to and attracted by experiments with welfarism carried out there. This created a deal of interest in Australia.

Mount Isa

Collins House may well have gone on to own Mount Isa too, but for the strained relationships between the key figures in the small world of Australian mining directorships. Conflict between Collins house directors and Mt Isa Mines executives, including W.S. Robinson and W.H. Corbould, resulted in Collins House staying away from the northwest Queensland operation.[25] Instead, Mount Isa relied on US capital – this was an early example of the growing influence of US investment after 1945.

Mount Isa Mines (MIM) was formed in 1924, and mining leases were consolidated in the late 1920s. From 1930, US capital was dominant, in the form of the American Smelting and Refining Company (ASARCO).[26] This meant that Mount Isa, which was backed by a strong and paternalist single company, became much more of a company town than Broken Hill.

Kambalda

The long reach of Collins House into the mining landscapes of Australia continued with the formation of Western Mining Corporation (WMC) in 1933, from a Collins House company, Gold Mines of Australia Ltd.[27] Gold Mines of Australia was formed in 1930 to develop goldmining operations; they were among the few mining operations that showed good prospects during the Great Depression. As West Australian properties were acquired, WMC was formed to manage them. WMC's chairman was the central Collins House figure, Sir Colin Fraser, and the board included W.S. Robinson and M.L. Baillieu. Head office was the famous 360 Collins Street, Melbourne, and there was a branch office in Perth (later moved to Kalgoorlie).[28] It was from the Melbourne office that directors planned mines and processing facilities, and followed the activities of competitors.

Such were the financial, technological and political factors shaping industrial and mining development in regional Australia. These towns were connected to, and controlled by, finance in the capital. But that is to privilege power over experience, and industry over history. We see a different picture when we move from a continent-wide focus to the towns themselves, their inhabitants, and the vibrant cultures established there.

Sirens, steam whistles and hooters

Industrialisation brought major changes to otherwise quiet country regions. The early development work, as pick and shovel broke the ground and felled trees held up temporary camps and shelters, evoked the old rough-and-ready pioneering days. This initial development phase is most commonly remembered in local museums and historical societies. Soon, however, residents watched in awe as poppet heads rose, industrial buildings were erected, railways were established, and industrial-scale mining began. Overhead baskets on aerial ropeways carried the ore from the pit top to the smelter with increasing efficiency. Mines and smelters worked around the clock, and gas and electric light glowed eerily from the mines.

This new industrial regime marked out the days with sirens, steam whistles and hooters, as the increasingly mechanised production processes disturbed the quiet of the landscape. These sounds and routines of industrial-scale mining replaced the seasonal rhythms of pastoralism and the deeper patterns and understandings of place of Indigenous Australia, of which most new white Australians were unaware. At the same time, emanations from chimney and furnace gave these sites a new odour and graced the atmosphere with particulates, smoke, sulphur dioxide, other gases and heavy metals. In dry country like Broken Hill or Mount Isa life was lived with a patina of dust that reached into every conceivable part of working and domestic life. Dust was, and still is, a common feature of life in Kambalda. In Queenstown it was water and the damp that pervaded everything. The Queen River and its associated creeks and springs filled the valley with the sound of moving water. It was this long-suffering river that carried away the worst of the Mount Lyell mine's pollutants, as general manager Robert Stitcht's pyritic smelters produced large volumes of sulphur dioxide – enough to further denude the magnificent hills.

The local labour markets expanded rapidly, often without even basic infrastructure like housing or a commercial sector. Large numbers of men arrived in town in the start-up phase, taking rooms in hotels or boarding houses or living in hastily constructed camps or company housing. The perspective of those who did not stay for long provides an important counterpoint to the memories of locals who lived life in one locality. Paul Said, for example, was a Maltese immigrant who arrived in Australia in 1924. After spending a year in Adelaide, he went to Broken Hill, where he worked and lived at the Grand Hotel. There were 134 Maltese-born people recorded in Broken Hill at the 1933 Census (106 males and 28 females). Like Said, many of them stayed in boarding houses or hotels. Their views reflected their experiences on the fringes of the labour market: they were, typically, exposed to the most dangerous work. According to Said, employment was good:

> [b]ut Broken Hill was a place you don't want to stay for too long. If you get a job you stay there for two or three years the most … two years is the most before the lead would get into you.[29]

Said later moved back to Adelaide and finally settled in Melbourne.

Mount Isa was similar to Broken Hill, with large numbers of workers moving through the local labour market and not necessarily staying. Max Thompson was one who moved from town to town. He was born in Lambing Flat in New South Wales in 1919. After war service (1939 to 1945), he worked at the coal mines in Lithgow, then on the Snowy River and Ord River schemes, then arrived in Mount Isa in the 1950s, where he worked at the concrete plant:

> We rented a – they call them 'flats' – but it was a two-storey shed. We paid thirty dollars a week for it … Thirty dollars! But you couldn't get accommodation.[30]

The global and national context

Rapid growth attracted large numbers of migrant workers, and many mining towns and industrial towns have a strong sense of themselves as crucibles of different nationalities. This can be expressed, and remembered, in a positive way. For example, there are specific ethnic-based clubs in Broken Hill, and there are Finnish churches in Mount Isa. These positive characteristics of multiculturalism sit uneasily alongside less happy experiences.[31]

The connections and flows of workers and their families are one set of experiences that turn our attention to the utilitarian demands of labour markets. These patterns need to be set against an internally directed process which saw many people develop attachments to their locality, often in a very short period of time. Local cultures and sensibilities were established quickly, despite the pace of change and the extent of population movement. Progress associations, churches, sporting associations, friendly societies and trade unions were all early arrivals, and their roots were set down quickly. These developments were inspired by both a growing commitment to the locality and an effort to address the very real problems that came with rapid growth. The lack of a water supply, formed and metalled streets, and a sanitary service joined less specific but no less important issues such as the absence of friends and the need for the company of like-minded people in the organisations of civil society. Municipal government usually arrived within five years of a major mineral find.

Taken together, the oral history interviews also reveal a deep connection to place in all the mining and industrial towns, from pre-existing townships to newly constructed, purpose-built company towns. All respondents were engaged in their town. They had knowledge and opinions they wished to share. They recalled their lives in vivid detail. Margaret Jones, for example, came to Mount Isa from Brisbane in 1957 and remembered her first house as a young married woman:

> It wasn't a flash house. It was a dirt floor. Tin walls and the windows were shutters that I pushed out ... Over the back of the hospital. No hot water. In summertime when it was so very hot the

pipeline was above the ground, so when it was very, very hot I had hot water in the pipeline [laughter].³²

Asked how one could describe Mount Isa to an outsider, Sigge Lampen paused, then replied earnestly:

> It's very hard ... It is just a mining town where people drink and get carried away, and work, and you know. Look, in this town there are so many nationalities. I think it would be hard to explain Mount Isa to anyone.³³

Their lives, and their memories, are not contained or defined by national, and cultural, assumptions about these places. Respondents often had relatives who moved from other mining towns. Elizabeth Firth's mother came from Kadina in 1917, one of the many South Australians who moved to 'The Hill' from the early 1880s on. Her father was born in Broken Hill, but his family was from the Victorian goldfields.³⁴ High rates of migration and the mix of nationalities also contradict the image of life in regional towns being stolid, monochromatic, the inverse of city cosmopolitanism. This is perhaps best exemplified by Sigge Lampen, a Finnish migrant who loves his Mount Isa life. Having lived in Finland, Sweden and Canada, as well as travelling extensively, he defines himself as a 'global citizen'.³⁵

In strong union towns the working class held great sway in political as well as cultural life. Brass bands, Eight Hour Day parades and labour-run newspapers were common in Newcastle, Broken Hill, Rockhampton and Townsville. The Victorian-era code of 'respectability' was dominant, but not uncontested. To middle-class reformers, it meant thrift, hard work and sobriety, and was often critical and judgmental of working-class culture. At the same time, however, 'respectability' was adopted by the working class and given new meanings: it meant not going on spending sprees or wasting resources, because these could create hardship for families, particularly when waged work was intermittent. In the hands of more radical, sometimes militant unionists, 'respectability' included the notion of the dignity of manual labour. It was used to argue for wage increases and better working conditions.[36]

'Respectability' also maintained the family and gender roles that were widely accepted in these towns. In the capital cities, aspects of the gender order were contested by first-wave feminists from the late 19th century on, but such debate rarely surfaced in the regional centres, much less in the more isolated towns. 'Respectability' meant a male breadwinner and a female housekeeper. This was given legal expression in the Harvester judgment in 1907. This was a Commonwealth Arbitration Court decision which said that a basic wage was defined as the amount a man needed to survive in a civilised community whilst supporting a wife and three children 'in frugal comfort'.[37]

From this a number of economic consequences followed. Men of all classes already had privileged access to the formal economy and the world of paid work, while women laboured in the town's homes or backyards in the informal economy. This became an ideal to which all classes aspired. Overall, women constituted 20 per cent of the paid workforce in 1900. However, their continued participation in that world was a necessity for many working-class households. In mining and industrial towns, though, there were fewer paid work opportunities for women. In the cities younger single women could find employment in manufacturing; in mining towns there was little beyond domestic service. All mining work, and most industrial work, was for men.

Underground mining in particular was strongly masculinised and was represented as both a heroic activity for 'real men' and as a dangerous occupation. Despite the increasing levels of mechanisation in the 20th century, mining and construction remained pick and shovel work; hard manual labour that taxed and tired the bodies of older men especially. This was true in all the case study towns until the late 1940s, when mechanical diggers, loaders and trucks began to replace manual work.

The struggle over occupational health and safety is a common theme in the mines, and in surface or smelter work. Rates of death and injury from mining employment were high. It fired union agitation and campaigns including the famous 'Big Strike' at Broken Hill in 1919–20, the Mount Morgan lockout in 1925 and the Mount Isa dispute of 1964–65. The Australian Workers' Union (AWU) was an important representative of rural workers, including miners, from the 1890s, and it came to dominate union representation at all sites except Broken Hill. At 'The Hill' a very particular union formation developed: a unique local alliance of workers that dealt with the Mine Managers' Association. This arrangement, which negotiated three-yearly agreements outside the state-sponsored arbitration system, survived from 1926 through to 1986.

Another common workplace experience was exposure to potentially harmful dust and fumes. This produced generations of men with 'miners' lung', silicosis and phthisis. Many of the men coughed and wheezed their way through oral history interviews, demonstrating the all too real legacy of these workplace hazards.

Lead is of special concern in metal mining and smelter towns. People first became aware of the effects of lead dust through workplace exposure. Port Pirie was an unfortunate pioneer in this regard, with lead poisoning the subject of union agitation and a government Royal Commission in the 1920s. When the Mount Isa lead smelters began production in 1931, workers in some parts of the operation inhaled lead-bearing dust. A government survey at Mount Isa in 1933 found that 27 per cent of the workforce suffered from 'mild lead poisoning'.[38] From the 1970s and 1980s on, the issue became a community-wide one, with a focus on the lead levels in the blood of children in particular. At Mount Isa, concern about this continues to be dealt with by companies, governments and the local community. Today's claims of lead poisoning and possible damage to yet more generations of Mount Isa's children are still sometimes greeted with the scepticism and surprise that have characterised industry and government responses since the 1920s. Exposure to environmental toxins may well, unfortunately, be another way in which these localities shape people.

Most of these towns were dominated first by migrants of British origin and then, increasingly, by locally born non-Aboriginals. There were still, however, pockets of non-Anglo groups and Indigenous people living in and around these developing centres. Italians in Port Pirie mostly came from Molfetta, a small village in southern Italy which had a tradition of extended trips away from home to earn a living. Not all chose to return and a community was formed. In 1933 there were 214 Italians out of a total population of 11,677. At Broken Hill there were significant groups of Italians, Yugoslavs, Greeks, Germans and 'Afghans'. Approximately 748 'foreign workers' out of a total workforce of 11,500 laboured on the line of lode in 1914. Of those, only 104 spoke English fluently.[39] Likewise, local storekeepers sometimes complained of small shops run by 'foreigners' opening after hours, and employing non-union labour.

Ethnicity could become a significant point of division within the local community and the local labour movement. This situation became more apparent after 1945 as larger numbers of migrants came to Australia. Finns, for example, were a significant part of the workforce at Mount Isa in the 1950s and 1960s. While there were strained relationships between Anglo and non-Anglo workers during the dispute of 1964–65, there were concerted attempts to overcome such divisions. Ethnic difference played out in a variety of ways and left a range of legacies, from the Afghan mosque on the northern edge of Broken Hill to the Finnish Lutheran and Pentecostal churches in Mount Isa.

After the upheaval of World War II, the world economy settled into 30 years of sustained growth. This underpinned strong wages growth and prosperity in many mining communities. Underground workers in places like Broken Hill and Mount Isa also enjoyed the benefits of the lead bonus, an over-award payment linked to the price of the base metal. In his detailed study of the lead bonus at Broken Hill, historian John Shields notes that between 1948 and 1966 workers were earning up to £165 per year in lead bonus, which represented as much as 50 per cent of their weekly wages. The figure rose and fell according to the lead prices, but in this period the bonus never went below £67 per year, or 16.3 per cent of total earnings; it peaked in the mid-1950s. It was a strong incentive for workers and their families to stay in Broken Hill, or move to the town. Elizabeth Firth married an accountant from one of the mining companies in 1956, at the height of the postwar boom. Although she had to leave her position in the office at a mining company when she got married, prospects were still good for the newly married couple. They set about building their own house. She recalls that 'we were quite happy to stay here because at that time the wages were very, very good. In the mid-fifties the lead bonus was extremely good and it was a way of saving.'[40]

The suburban ideal, which flowered in these towns in the post-1945 period (between 1945 and 1972), was celebrated with great zeal. The achievement of a 'normal' suburban existence in harsh inland locations represented a modernist taming of a wild environment. It was underpinned by wages that were high for working-class families, which helped offset the generally high cost of living. However, the new suburbs never quite matched the ideal. The mines or smelters were never far away, and they were an ever-present reality that might destroy the apparent peace of outback suburbia. Kambalda represents the strongest expression of the designed suburban ideal, an uneasy mix of corporate paternalism and carefully targeted control.

While Broken Hill, Mount Isa and Port Pirie continued to grow in the 1960s, and Kambalda was born in that decade, Mount Morgan and Queenstown were already suffering as their ore bodies came to the end of their productive life. This led to mine closures in the 1980s. Even the fabulous 'Silver City' of Broken Hill declined from the 1970s, and urgent efforts to diversify there, as elsewhere, started a new phase of existence. The dreams of great growth and progress always contained a flicker of doubt in these towns. Such doubts are considered in more detail in later chapters.

These towns were places where international trade, migration, and the influence of state and federal governments met local environment and culture. They are permeated with capitalist and modernist ideals of industry, progress and profit. Money oils the wheels and civil society cements the foundations of these towns. These places were not simple nodes of capitalist production, barren and cultureless centres. They were vibrant and contradictory places, enmeshed in links

The global and national context

to nearby towns and capital cities. At key points these towns took root, establishing, if they were lucky, a life that made them more than workplaces. In the following chapters we turn our attention to the six towns.

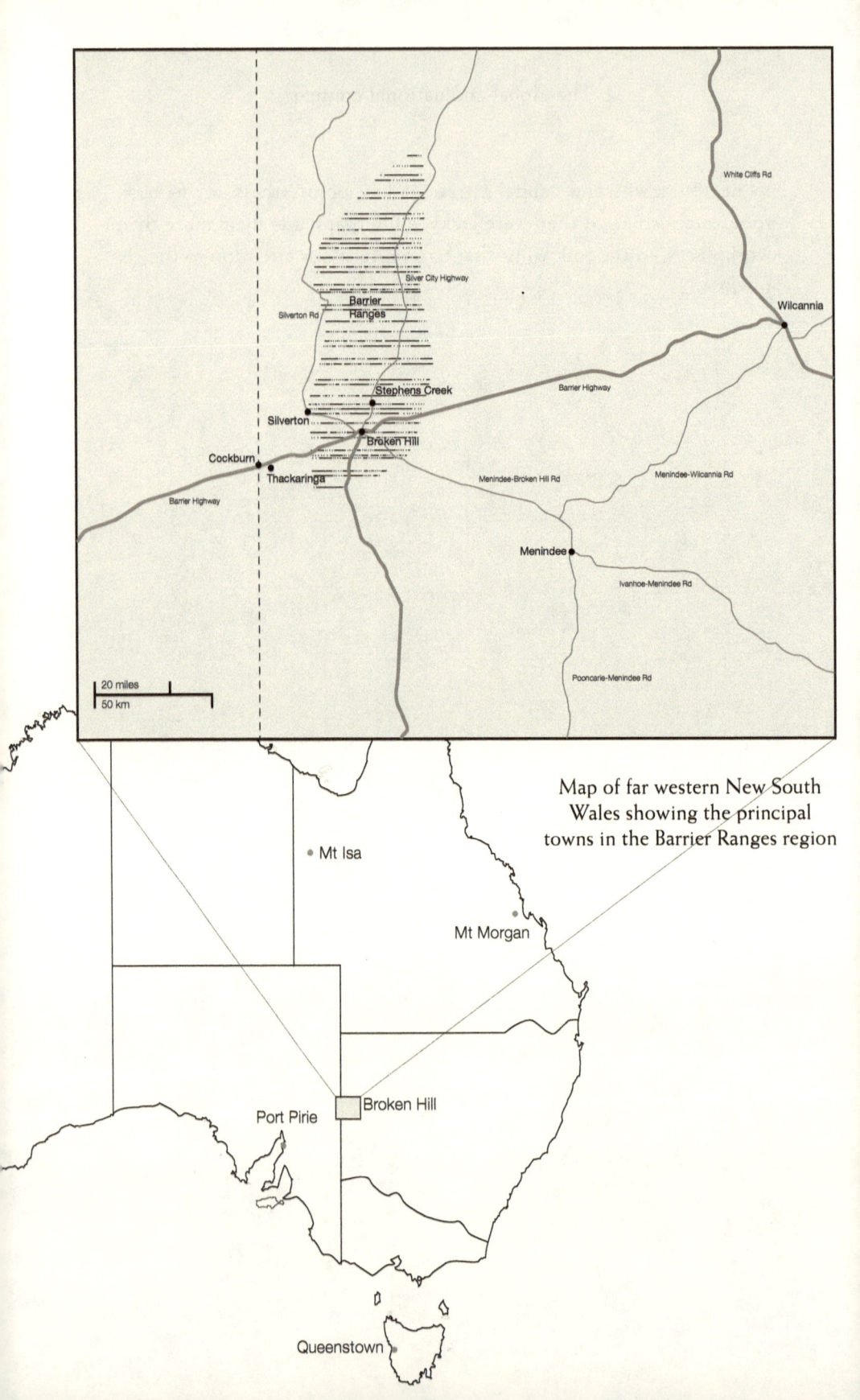

Map of far western New South Wales showing the principal towns in the Barrier Ranges region

2
BROKEN HILL
Icon of working-class culture

On a warm April day in far western New South Wales, a sharp blue sky frames an outback city. A breeze graces this quiet Easter Monday, as cars meander up and down the main street. By the afternoon, the wind has warmed, and blows piles of leaves about the streets, up and over the high gutters. Clouds east of the Darling River have built up, and are now a looming presence on the distant horizon. Later, these clouds will bring rain, bathing the Silver City in a few glorious moments of relief; a cleansing wash as heavy as it is brief.

In this dry country, a distinctive ridge 300 metres above sea level became an iconic landmark. With the discovery of rich silver-lead-zinc bearing ore in 1883, the town of Broken Hill grew. The 'Broken Hill' is now notably lower, replaced by mullock heaps and tailings. Its colours range from desert ochre to basalt grey to slag heap black. The town's grid-pattern streets and clearly marked boundaries impose an odd kind of order on the rolling reddish hills of the ancient ranges. The mines are now mostly silent – as is the main street on this sleepy public holiday – but the town is alive to the possibilities of a post-mining future.

If there is an exception to the general cultural marginalisation of mining and industrial towns, it is Broken Hill. One important and obvious advantage is its size. A city with a population of 20,139 people in 2006 (according to the Australian Bureau of Statistics) is difficult to ignore. There has been a decline since the halcyon days of the 1950s and 1960s, when the town's population edged over 30,000, but Broken Hill has not experienced the steep decline of others, such as Mount

Morgan. Broken Hill also has a cultural presence. Academic and popular works have covered various aspects of its history, and the town has a thriving family and local history scene with its own publishing profile.[1] Also, it has appeared in numerous works of literary and popular fiction and in films.[2]

Despite the fact that Broken Hill sits squarely in the national consciousness, that consciousness does not recognise the complexity of the town's history and culture. Broken Hill is the 'Gibraltar of unionism', a site of industrial conflicts, the origin of major multinational mining companies, an old mining town with residual working-class toughness. It figures in national and state histories as a site of economic development and industrial conflict, typically mentioned for the discovery of the orebody in 1883, the rapid development of the town, and the 'Big Strike' of 1919–20. Many histories, both scholarly and popular, focus on the period before 1920 and end with the Big Strike, as though history finished with Broken Hill then.[3] Yet it has a diverse and multicultural history, full of class, religious and cultural complexities.

The region

Broken Hill sits so clearly and starkly in its arid environment that its links with the surrounding region can be easily overlooked. The town now lies at the crossroads of a number of regional highways. The road from the west is through South Australia, from Yunta through the old railway town of Cockburn at the SA/NSW border. The country is flat and scrubby. Author and diplomat Paul McGuire described the country between Peterborough and Broken Hill in the late 1930s: 'Here are the saltbush and the country from which the saltbush is gone, tracts of "scalded" earth, stripped down to the clay.'[4] The dust storms come out of this overgrazed pastoral country. On average Broken Hill has just over 30 days of temperatures over 30°C, and 6 days per year over 40°C. The annual rainfall is a modest 258mm, while June, July and August can be chilly, with average maximum temperature between 11°C and 13°C.

Broken Hill

From the south the Silver City Highway travels through Mildura and Wentworth to Broken Hill, then continues through to Tibooburra in the far northwest of New South Wales. From the east, the Barrier Highway passes through Cobar and Wilcannia, bypassing Menindee, which lies on the Darling River to the southeast. Another access is via the NSW rail line, which runs through Darnick, closer to Menindee.

That's the situation today. In the middle of the 19th century, the region was on the margins of European occupation. From Sydney it was the 'far west' – the term described both distance and direction. Explorer Charles Sturt journeyed from Adelaide through the region in 1844, naming the ranges 'Stanley's Barrier Ranges' after Lord Stanley, at that time secretary of state for the colonies.[5] Sturt made no specific reference to a 'Broken Hill', despite numerous claims in the secondary sources to this effect.[6] The name 'Broken Hill' comes from a later pastoral era, when the area was part of the Mount Gipps run and was known as the 'Broken Hill paddock'.[7] The Darling River, the path of Sturt's journey from Adelaide, developed as a major trade and transport route into the area, supplying the emerging pastoral industry and transporting the wool to market through the river ports.

The goldrushes of the 1850s and 1860s generated great interest in mineral wealth and prospecting. Throughout the 19th century, men wandered the back country searching for gold in isolated creeks and gullies, and the far west of New South Wales was no exception. For example, a silver-lead mine was established at Thackaringa, west of Broken Hill, in 1876. A better find at Silverton soon followed. In 1880, a mining registrar was appointed and a town developed at Silverton.

The discovery of rich ore in the 'Broken Hill paddock' by Charles Rasp in 1883, and the eventual formation of BHP has been covered in Chapter 1. BHP dominated, but the selling of the southern and northern ends of the mining lease made the town a multi-company site. The multiple mines and the rise of smelting along the line of lode gave the labour market diversity and size, and it soon eclipsed Silverton. The population at Broken Hill had reached 19,789 by 1891.[8]

The town's first ten years, from 1883 to 1893, were characterised by competition and tension with Silverton. Silverton was proclaimed in 1883 and by 1885 boasted 5000 people, but by the beginning of the 20th century the population had declined to 1901; the municipal council had been withdrawn two years earlier. Throughout the 1890s, bullock teams plied the 25km between Silverton and Broken Hill moving wood, iron and brick buildings piece by piece, every item a symbol of the vanished dreams of Silverton's greatness. Broken Hill's final victory was its widespread adoption of the 'Silver City' tag, dwarfing 'Silverton' both in name and in reality.

The links between the pastoral country surrounding Broken Hill and the town itself are strong. In 1907, the West Darling branch of the Pastoralists' Association was formed – it held its first meeting in Broken Hill. The major pastoral stations in the area were Mount Gipps, Kinchega, Mootwingie (now known as Mootawingee), Corona, Mundi Mundi and Poolamacca.[9]

Oral history also reveals such connections. One respondent who is very much aware of the links between the town and the pastoral economy is May Watson. Born in Broken Hill in 1942, she grew up on a 50,000 acre station about 45km to the southwest. She lived on the farm until she was eleven, being home-schooled by her mother. Then she boarded in Broken Hill with a number of other country children.[10] Watson's family, like many in the pastoral regions surrounding Broken Hill, relied on the town for supplies, services, and education. Even residents from far away White Cliffs (an opal mining town) travelled to Broken Hill to shop.[11]

Making a town

By 1885, Broken Hill had three hotels and a mining camp. The town was surveyed in 1886, and a municipal authority established in 1888. The main street, Argent Street, ran parallel to the line of lode, and the surrounding streets in the central area were named after minerals, or chemical compounds. The line of lode cut the growing town in two, with employees from Broken Hill South P/L living on the southern side of the ridge, an area that became known as South Broken Hill or Alma.[12] The thick mulga scrub that covered and surrounded the actual Broken Hill disappeared through a combination of overgrazing, tree felling (for mine props, furnaces and home fires), and the effects of large amounts of mine spoilage being dumped there. One pioneer who was on site in 1883 recalled that 'The spot where Argent street now is was thickly timbered with mulga. I once shot an old man kangaroo just about where Finn's Hotel stood.'[13] Photographs from the period confirm the wooded, bushy appearance of Broken Hill.[14] By the early 20th century, dust storms were common, and the few surviving trees fringed creek beds.[15]

But it was always the line of lode, dividing the town in two, that dominated. It became a great elevated swathe of dusty mines and industrial workplaces tracking northeast to southwest, with more than a dozen chimneys from the smelters belching smoke and steam, criss-crossed by aerial ropeways and rail tracks transporting ore. Nothing grew along the line of lode bar the hardiest weeds. Alongside this industrial complex, Argent Street developed as a lively and sociable counterpoint, with Saturday afternoons full of commerce and interaction.

The line of lode

By the 1880s, mining and pastoralism were the dominant activities in this region, displacing the Indigenous economy and way of life. After Rasp's discovery and the establishment of the Syndicate of Seven which underpinned the formation of BHP, a town developed rapidly. Without the mines, there would have been no Broken Hill.

Individuals, syndicates and larger companies took up mineral leases in the surrounding areas. The fabulous wealth of the BHP mine attracted great speculative activity. Would the surrounding leases be as rich? Many of the smaller leases were not viable or were soon sold to larger companies. By 1893, there were nine major mines and a number of mills and smelters, but the line of lode was still dominated by BHP, 'the big mine'.[16] By 1909, BHP employed 2815 of the town's approximately 8000 mine workers. By then most workers at BHP and the other mines were wage earners in large industrial concerns, rather than hopeful partners in small ventures.

In 1911 there were 7972 workers engaged in mining and quarrying. The mines were complex operations, requiring not just a miners, but also workers in construction, smelting, transport, engineering and general mechanical repair. There were 4146 males listed in the 'mining' industry at the 1921 Census, but also 934 in 'manufacturing', 377 in 'transport and communication', and 186 in 'building and construction'. Population growth at Broken Hill, together with its isolation and the high costs of transport, made sure it had a vibrant commercial and service sector. At the same Census, there were 886 male workers in the 'commerce and finance' sector, 348 in 'professional and administrative' work, and 194 in 'personal and domestic service'. The commercial and service sector was particularly important for female workers. Of the 1381 females who listed an occupation, 538 were in the 'domestic' category, 327 in 'professional' and 325 in 'commerce'. Broken Hill was not all about mining and miners, though they dominated, then as now.

Mining was a particularly dangerous and difficult occupation. The Broken Hill lode was subject to 'creep', as noted in Chapter 1, which forced many of the smelters to relocate to other sites. Frank Bartley, a miner at the Central Mine, recalled what happened when the ground moved and subsided along the line of lode:

> none of them belts [on the mills] would keep on when it went down ... In about two hours, the ground had a great crack ... When it went down it showed a great big crack, and that crack opened up so that you couldn't jump across it.[17]

Early labour movement historian George Dale quotes Mines Department figures indicating that between 1883 and 1917, 390 men died on the line of lode, and 729 had serious injuries.[18] A Communist Party publication suggests that 501 men lost their lives on the line of lode between 1885 and 1932. As the author notes, these figures do not include the 'thousands of men who worked until work became a physical impossibility, and then coughed away their lives away with tuberculosis'.[19] Square set timbering was introduced in the 1880s, but was increasingly replaced by open stope mining, where miners worked into the ore face, blasting and drilling ore from an ever expanding cavity which was usually self-supporting. Without support from timber bracing, however, an open stope was prone to cave-ins.

One such accident occurred in 1919 when George Boucher, a single man from Victoria with 30 years' mining experience, was knocked to the ground in the North mine at the 1250 feet level (380 metres) by a 250kg piece of rock that broke away while he was in the stope. The support timbering was three or four sets away (4 metres), but the shift bosses and other miners had deemed the stope safe. These miners were working under contract, so there was a financial incentive to maximise production. Boucher died later that day in Broken Hill hospital. The jury returned a verdict of accidental death but recommended that the square sets or supporting timbering should be extended closer to the stope.[20]

This fatality epitomised the major concerns of the Amalgamated Miners' Association (AMA) over safety: open stoping, and a form of payment that encouraged men to take risks. Underground conditions were hot, but ventilation through the complex of shafts, drives and stopes could create draughty and cold conditions, so many miners were vulnerable to pneumonia, exacerbated by dust. The dust created other respiratory problems: 'miners' lung', or phthisis. Miners and smelter workers ingested lead-bearing dust or inhaled leaded fumes, so lead poisoning was another common problem. It's little wonder the unions and the friendly societies provided sickness and accident insurance. These conditions, together with wage rates, fired union campaigns in the early 20th century. The Great Strike of 1919–20 led to the *Workmen's Compensation (Broken Hill) Act* of 1920 and the *Workmen's Compensation (Lead Poisoning – Broken Hill) Act* of 1922.[21] Such victories were only achieved because Percy Brookfield, Broken Hill's legendary Industrial Labour Party member of the Legislative Assembly, held the balance of power in the NSW parliament until his untimely death in 1921.[22]

Taking root

The initial work of developing a new society in far western New South Wales was done at Silverton. Those pioneering individuals and organisations increasingly relocated to Broken Hill after 1885. A Progress Committee was established at Silverton in early 1884. The major churches had a presence in the town, and workers established the Barrier Ranges Miners' Association in 1884. The hotels at Silverton played host to these new organisations. The first meeting of the Miners' Association was held at John de Baun's Silverton Hotel.[23] The proximity of South Australia and the decline of its copper mines meant that large numbers of South Australian miners and their families came to Silverton and (later) Broken Hill. Such migrations were common over many generations. Frank Bartley, for example, was born in 1888 in Moonta, but was taken to Broken Hill by his parents when he was just two weeks old. His father was from Cornwall and they had come to Broken Hill, like so many Cornish miners, or 'Cousin Jacks', via Burra and Moonta.[24] Elizabeth Firth's family on her mother's side arrived in Broken Hill in 1915 from the South Australian copper mining town of Kadina on the Eyre Peninsula. Rapidly growing mining towns such as Broken Hill were populated by those who had already migrated once, and were now moving again, often from declining mining towns. Putting down roots was not an easy task in their world.

William Delamore, a Silverton publican, relocated to Broken Hill in July 1885, thus becoming the first hotelier, and owner of the first building of any substance, there. Delamore's Bonanza Hotel was transported from Purnamoota, a silver mining settlement north of Silverton.[25] The hotel, despite its formal name, was known by locals as Dellys, and included a boarding house, which was run by Mrs Delamore until her death in 1893. These boarding houses were crucial institutions in mining towns, as temporary accommodation was in high demand. By 1888, there were an incredible 55 hotels in Broken Hill, with most offering accommodation. There were also a number of boarding houses, large and small.

Delamore was typical of the early storekeepers. He served on the Broken Hill Progress Committee in the late 1880s, was Alderman on the municipal council in the 1890s, and dabbled in horse racing. Opposite Delly's was Sully's store, also moved from Purnamoota and opened in 1885. All these premises were on mining leases, and before the official survey of 1886 they had no security of tenure.

The Broken Hill Progress Committee first met in late 1885, focusing on lobbying the NSW government for new services and facilities, including a postal service, a hospital, a reliable water supply, new roads and a railway connection. Complaints about the water supply persisted for decades, as the town kept outgrowing private – and later government – attempts to secure a reliable service.

The private water supply and the privately owned railway service were symbolic of Sydney's lack of interest: the NSW government seemed loath to invest for fear that Broken Hill would go the way of other short-lived mining settlements. This allowed space for private enterprise. The Broken Hill Water Supply Company, formed in 1890, emerged as the winner out of a collection of competing syndicates.[26] The company developed the Stephens Creek reservoir and delivered the service until 1916. The railway to the South Australian line at Cockburn on the SA/NSW border was built by the Silverton Tramway Company. It first linked the South Australian system to Silverton and was later extended to Broken Hill. The line terminated at the Sulphide Street Station. The company was a major employer, running over twenty locomotives in the 1930s; it was one of the few profitable privately owned railway lines in Australia. The NSW government line from Parkes via Condobolin and Darnick arrived in Broken Hill in 1927. The Silverton line was closed in 1970 after the extension of the standard gauge line from Cockburn to the NSW system, with the old station becoming a railway museum. There was so much growth in Broken Hill that a government-run tram system was eventually built. The trams, which were old steam-driven vehicles transferred from Sydney, operated in all the suburban areas of the town from 1902 until 1926.[27]

Religious organisations and friendly societies were quick to establish congregations and branches in Broken Hill. They gave new arrivals a ready-made community, offering both material support and fellowship. The first organisations at Broken Hill arrived by way of Silverton. A Catholic Church was completed at Silverton in 1886, but even in 1885 mass was being held in Broken Hill, and the first of many Catholic churches was built in the town in 1887. The Bishop of Wilcannia (which included Broken Hill) was Bishop John Dunne, who was a strong and vocal supporter of the miners during the industrial disputes of 1889 and 1892.[28]

The Catholic community also brought friendly societies such as the Hibernian Catholic Benefit Society, formed in 1887. The Hibernians offered sickness and accident insurance, the fellowship of an organised and active branch structure which supported the Church, the Catholic Club (which opened in 1912), and special events such as St Patrick's Day.[29] By 1934 the friendly societies, including the larger organisations such as the Manchester Unity Independent Order of Oddfellows and the Independent Order of Oddfellows, had some 4800 members on the books, which represented approximately 40 per cent of Broken Hill's total employed population.[30]

Early union organisation was underpinned by a belief that masters and men had shared interests. Historian Brian Kennedy suggests that this was a result of the difficult conditions on the line of lode and the generally democratic spirit of the silver mining boom.[31] Associations were characterised by cross-class membership: workers and employers all joined. Broader changes throughout the colonies slowly undermined this alliance and class became a stronger organising principle. By the 1890s most miners were wage earners or contractors working for a joint stock company with absentee owners and on-site managers, and by 1892, the managers had organised separately through the Barrier Ranges' Mining Companies Association, mirroring the behaviour of ship owners, pastoralists and coal proprietors throughout the colonies.[32] This organisation later became the Mine Managers' Association (MMA).

The clearest expression of this class mobilisation was a famous series of strikes and lockouts which began in the late 1880s, punctuated the succeeding decades, and culminated in the epic Big Strike, which ran from 1919 to 1920. None of this can be separated from the town's sense of isolation and neglect. The NSW government was slow to respond to its demands. As noted, water crises persisted well into the 1950s, until the Menindee system was linked to the water supply. Grievances were exacerbated by the high cost of living, itself a result of high transport costs. Evidence presented to the Commonwealth Arbitration Court indicates that wholesale prices in 1909 were 11 per cent above Adelaide prices, and retail prices for items such as beef steak, corn, mutton and cabbages were almost 50 per cent higher than city prices.[33] The earlier cross-class connections had reflected a sense of shared hardship, but by 1920 trade unions spoke for Broken Hill.

The 1909 dispute, which began after BHP announced a 12.5 per cent pay cut, helped establish an image of the town as a bastion of union solidarity. Photographs of the miners and their supporters marching behind a banner proclaiming 'Behold, the workers think' were widely distributed in press reports and on postcards. Such images attracted socialist and syndicalist activists such as John Joseph O'Reilly, Mick Considine, Percy Brookfield and Torliev Hytten to the town, adding to the core of home-grown activists.[34] There were some setbacks, including the gaoling of union leaders and the use of armed police in 1909, but there were also some major victories: after the Big Strike, miners won access to a workers' compensation scheme, a 35-hour week and new safety measures, including mandatory wet drilling. Broken Hill became an icon of working-class organisation; this image captured the increasing power of the organised working class but obscured the complexities of the town's social and political life. The Labor Party controlled the municipal council from 1900 onwards, in one of the party's earliest and most successful moves into local government in Australia, and this further solidified the town's working-class image.[35]

But internal divisions were powerful. They were based on religious affiliation, schisms in the labour movement and even geography.

Internal conflict in the labour movement occasionally reached epic proportions, and was as bitter and as earnest as any fight against the heartless capitalists. The battle against a 'bogus union' – the Barrier Workers' Association, which had close links with the mine managers – raged throughout the late 1910s. Other unions, such as the Federated Engine Drivers and Firemen's Association and the AMA (which changed its name to the Workers' Industrial Union of Australia [WIUA] in 1921), along with 'town sector' unions such as the Municipal Employees' Union, were a volatile mix of traditions, strategies and political orientations. Each was internally factionalised. Such divisions plagued early attempts to build peak bodies with all local unions represented. But in 1923 the Barrier Industrial Council (BIC) emerged. It overcame internal differences and influenced generations of local workers.[36]

The connections between religious affiliation and the labour movement were intricate. Even though the Catholic Church was a vocal supporter of the miners, early labour movement leaders were overwhelmingly non-conformist Protestants influenced by the Wesleyans, the Congregationalists and the Primitive Methodists. The Cornish miners who relocated from Moonta, Kadina and Wallaroo brought with them sober, respectable and well-organised union, church and friendly society groups.[37] John Henry Cann's political career is typical of many of the early labour leaders. English-born Cann was a former Primitive Methodist lay preacher. He moved through the coalmining communities of New South Wales, where Primitive Methodism was strongly associated with working-class communities, and relocated to Broken Hill in 1889, working at the BHP mine. By 1890 he was president of the local branch of the AMA, and active in the local Congregationalist church. He and his wife were active members of the temperance lodge, the Independent Order of Good Templars.[38] In 1891 he became the first Labor member for Sturt, a seat he held until 1913.[39]

Labour leaders of Catholic origin, such as Mick Considine and E.P. (Paddy) O'Neil, dominated from the 1920s. O'Neil was President of the BIC from 1923 until 1949. More than any single labour leader, with the possible exception of the much revered Percy Brookfield, he personified the Broken Hill labour movement. O'Neil had worked for the municipal council since 1913, and was an official with the Municipal Employees' Union. In the early 1950s, he was also active in the Broken Hill branch of the Campion Society, a Catholic intellectual movement which encouraged members to apply their Catholicism in everyday life. His biographers describe his views as 'socially conservative'.[40] The Catholic-inspired anti-communist forces, known as the Groupers, were active in the local labour movement in the 1950s, as were left-wing groups including communists. Candidates of Catholic origin did well. L.A. 'Lew' Johnstone was the successful Labor candidate for the seat of Cobar in 1965, transferring to Broken Hill in 1968. As historians Bradon Ellem, John Shields and Julie Kimber note, 'Catholic, anti-communist, Laborist and an ardent localist, Johnstone typified the world-view of the postwar WIUA leadership.'[41] Continuing the Catholic influence were W.S. (Shorty) O'Neil (no relation to Paddy O'Neil), who was BIC President from 1957 to 1969, and his son W.S. (Bill) O'Neil (BIC President, 1985–95).[42]

North Broken Hill was widely known as the 'Catholic mine' and the North Broken Hill suburb next to the mine was a Catholic stronghold, the site of the Sts Peter and Paul Church and the North Broken Hill Convent School.[43] Incidentally, Lew Johnstone worked at the 'Catholic mine' before moving into state politics, as did Shorty O'Neil.

Labour and locality

By the 1920s, 'local' increasingly meant settled, unionised and family-focused; all values that a range of union, civic and religious bodies were keen to support. Until the end of World War I, there had been many itinerant men coming to town. They were not union members, and when a fully unionised workforce (at least in the mines) was achieved in 1918, many of these single men drifted away. Broken Hill's population of females and males slowly equalised.

Married men were more militant, perhaps because they had more to lose.[44] They were connected to local organisations and kin networks that moderated the harsh working conditions and provided civic, sporting and recreational activities. However, these people and organisations could turn against a recalcitrant individual. The hatred reserved for non-union workers in the 1909 and 1916 disputes, powerfully revealed by the 'mock graves' of well-known scabs, shows that the community could use intimidation and the threat of violence to secure allegiance. This fusion of labour and locality was given institutional expression in 1931: the BIC closed its books to outsiders in an attempt to regulate labour flows, support the local community and provide jobs for local men first.

So while Broken Hill was increasingly thought of as a united working-class town, geography, suburban development and municipal politics complicated the picture. Suburban expansion, with its small corner shops and hotels, underpinned new forms of political organisation. Four distinct suburban areas emerged in the 1890s. By 1898, the police Census estimated that there were 11,387 residents in Central Broken Hill, 4546 in North Broken Hill, 4242 in Railway Town and 3902 in South Broken Hill.[45] There was a church and sometimes a school in each of these suburbs.[46] As distinct suburbs developed, the old Progress Committee was joined by the Broken Hill South Progress Committee and the Broken Hill West Progress Committee. These organisations focused on lobbying the council for better services. At the formation meeting of the West Progress Committee, chairman Frederick Brooks (a guard with the Silverton Tramways) noted that:

> the ratepayer of the West must recognise that they were an important body, for they belonged to the most important part of suburban Broken Hill. The object of the progress committee was to see that the wants of the West were attended to by the council, and a programme setting out those requirements would be at once prepared and put forward by the committee.[47]

These divisions were formalised through sport. The favoured male sport in Broken Hill was Australian Rules – another example of South

Australian and Victorian influences. In the 1880s and 1890s there were teams of 'South Australian' and 'Victorian' players, or mine-based teams such as 'Central Mine' or 'North Mine'. By the late 1890s new and more permanent club structures emerged, based on suburbs. The North Broken Hill club was formed in 1899, and Central Broken Hill, West Broken Hill and South Broken Hill (or Alma) in 1900. Local rivalry was fierce. In one infamous game in 1926 when Wests met Norths at Jubilee Oval, a group of 500 stormed the ground after a Wests player had been knocked unconscious. The brawl that followed resulted in 11 Wests players being charged with riotous behaviour; they received fines ranging from 10 shillings to £2.[48]

These sporting and political organisations reflected and reinforced local loyalties. As Peter Bennett recalled, the main town centre was called 'over the hill' by those in South Broken Hill. He also commented on the local football rivalries:

> Yeah, we love to beat the North. The Souths like to beat the North. Ah Westies... yeah, we beat them, but you know, the Westies are alright. They're Westies, you know. Centrals, well, you know, nearly as bad as the North but not quite.[49]

There were still opportunities, however, to represent the entire town. A combined Broken Hill team usually toured South Australia at the end of the season; season-long rivalries were then forgotten.[50]

Exclusions

Beyond the internal differentiation of Broken Hill society, there were divisions that were more fundamental. The Indigenous traditional owners of this country are the Bulali, part of the Wiljakali people. The Indigenous name for the hill itself is Willyama. The Barkindji people from Menindee Lakes also had an important relationship to the Willyama area. The Darling River and the Stephens Creek areas were the focus of traditional life. The extension of the European pastoral economy into this area was thus a direct threat to their food sources. After a period of conflict, some Aboriginal people found work in the pastoral industry. Often this was the only viable option for survival. After the passing of the *Aborigines Protection Act 1909* (NSW), many Indigenous people in the region were forced to relocate to missions or reserves. This was a common experience in all the Australian states from the 1890s onwards.

An Aboriginal camp site was reportedly on the outskirts of the new town in 1885, on the rise later 'occupied by the municipal crushing plant', but it is not clear whether this was a traditional camp site or a fringe camp – these were common on the edges of Australian country towns.[51] By 1898, the Broken Hill police estimated that there were 44 Aborigines in the area. There were sometimes incidents of racist violence. Sometimes groups of Aboriginal men came to town, to perform corroborees 'with a view of raising funds to provide them with the means of combating the coming winter'. One such event was reportedly harassed by a group of about 200 people – according to the *Herald*, mainly 'boys' and 'larrikins' – who 'commenced to shout and throw missiles'. The crowd had to be dispersed by the police.[52] This kind of attempt to raise money may have been a result of pastoral failures in the 1890s, which created hardship for all workers in this industry.[53]

Aboriginal men continued to work for the major pastoral stations, moving in and out of Broken Hill according to demand. Many of these workers, such as Indigenous Elder George Dutton, recalled visits to Broken Hill, even periods of residence there. Dutton had a cousin, Walter Newtown, who lived in Broken Hill. Some Aboriginal men, like Newtown, worked in the mines or for the municipal council after 1945.[54] Seasonal work fruit picking in the south, around Mildura, was another option.

In the post-1945 period there was a consolidation of the local Aboriginal population in Menindee and Wilcannia. The Menindee mission, which had opened in 1933, was closed in 1948, with residents required to move east into Wiradjuri country. Many refused, and instead relocated to Wilcannia.⁵⁵ There were also Aboriginal people at Silverton in this period, taking advantage of the cheaper rents there. Silverton was close enough to Broken Hill for paid work and entertainment. There was also a close connection between the Silverton Methodist Church and local Aboriginal people.⁵⁶ There were other supporters too: by the 1950s, the local labour movement was lobbying on behalf of Aboriginal people. The BIC, for example, moved against segregated hotels, pressuring hoteliers to allow Aboriginal men into the bars.⁵⁷

Continual forced movements or relocations compromised Indigenous relationships to new settler towns; non-Indigenous workers and their families were similarly affected. The views of those who did not live for long in Broken Hill are difficult to uncover in the historical record, and obviously cannot be found in these towns now, since they, by definition, moved on.

Rapid growth attracted large numbers of migrant workers. In 1933, there were 134 Maltese-born people in Broken Hill. Many short-term migrants stayed in the town's hotels, returning to Adelaide or Melbourne after a short time.⁵⁸ Other groups had mining experience, such as the Germans, or were able to fill a particular niche in the local economy. Census figures indicate that 30 'Afghans' and 38 'Syrians' lived in Broken Hill in 1921, most likely around the North Broken Hill area, where a well-known 'camel camp' had been established and where the Broken Hill Mosque was built in 1887. The 'Afghans' played a crucial role, with their camel trains linking isolated towns with inland and coastal supply routes.

Broken Hill has a strong sense of itself as a crucible of different nationalities. This is seen as a positive, and migrant history has a strong presence. However, it sits uneasily alongside the fact that migrant workers were always on the fringes of the labour market and allocated the hardest, most dangerous work.⁵⁹

Yet there is also evidence that deep local connections were felt not only by the locally-born, or the long-term residents. Harry Finlay's story highlights this. Finlay only settled in Broken Hill in 2002. He came to Australia from New Zealand as a young man in 1962. After a number of moves, he settled in the opal mining town of White Cliffs in 1980 and stayed for 22 years. Approaching old age encouraged him to move to Broken Hill. Moving from small to large regional towns is a common feature of rural demographics, particularly for the elderly. Interviewed in 2006 and asked, 'Can you tell me about what you've discovered in the town? What are your impressions of the community?', Harry replied:

> Oh ... it's a wonderful close knit [town] in one way and yet they're so open. I thought that White Cliffs was friendly but this is, I'd say, one of the most friendly towns that I've ever come across. It's like the old days [when the] old union had it all tied up and everybody working and pulled together. It's beautiful. And there's so much going on here and especially in the old mines and stuff like that, you know.[60]

Many people arrived with allegiances and links elsewhere. Local society, in particular the civic and voluntary sector, was a vigorous force encouraging participation and commitment.

Nonetheless, there were fault lines of difference and conflict. The BIC preference clause for local men obscured the exclusion of 'foreign' workers in the late 1920s. Moves against foreign workers, most of whom were Southern Europeans, were common, adding a racial element to the general suspicion of itinerant single men. Recent work by historian Sarah Gregson has underlined the importance of the mining companies, local clergy and conservative political groups such as the National Party in encouraging racism. Her analysis of the press battle between the conservative *Barrier Miner* and the labour-owned and controlled *Barrier Daily Truth* reveals a society with strong class and socio-economic divisions.[61]

This increasing localist and racial sentiment, and the BIC's efforts to control the labour market, are reflected in the Census figures. In 1911, 11.6 per cent of Broken Hill residents had been born in Europe.

By 1933 this figure had declined to 8.8 per cent. Other factors, including the continued weak state of the local mining industry, the effects of the war and the internment of Italians, Germans and other 'enemy' subjects, which removed them from where they lived, played a role too. By 1947, only 5.4 per cent of Broken Hill residents had been born in Europe.

Arriving

In the early days, most population growth was through new arrivals. Despite the exclusions and limitations noted above, most arrivals found a place; their integration was smoothed by the network of kin, religious, civic and political organisations.

Memories of arrival give a flavour of the town; the first glance of something new providing lifelong, telling images of difference. Clem Davison, who came by train from Adelaide when he was nine, in 1906, recalled arriving early in the morning at Broken Hill:

> We were naturally interested in seeing where we were going, but it wasn't till we were a few miles out that we could see the smoke coming from the stacks of the mines ... The roads were dirt and metal and there were very few trees. Altogether, it wasn't very attractive.[62]

The labour market expanded rapidly. Large numbers of men arrived in the initial growth phase between 1883 and 1900. Population growth outstripped the water supply and the dry, hot summers in overcrowded unsanitary conditions produced a typhoid outbreak, known as 'Barrier Fever'. In the late 1880s, the mortality rate at Broken Hill increased dramatically as the urban infrastructure failed to keep pace with population growth.

The male domination of mining and its supporting trades meant that there were more men than women in these towns as they grew. This was the case in all the towns we cover. In 1911, out of a total population 30,972 in the Broken Hill local government area, there were 16,921 males and 14,051 females. This imbalance had predictable and familiar consequences. As mining declined after 1970, and as working-class men on average died younger than women, the proportions of men and women changed, until women outnumbered men. By 1991, there were 11,667 males and 12,072 females in Broken Hill. The age of the miner had truly ended.

Making a town – postwar

After World War II, a settled compact between capital and labour, together with virtually continuous economic growth, provided a prosperous life for many. Demand for Broken Hill ore continued unabated. At the 1947 Census, the population for the Broken Hill local government area was 27,054, with only 356 more males than females. There were 5073 workers classified as 'mining'. By 1954, the population had risen to 31,351. Growth was mainly in the male population, as labour demand and the lead bonus were attracting men. By the 1954 Census, there were 1063 more males than females in the general population, and the numbers of workers in the mining industry had increased to 6175.

This was the peak of mining employment in the postwar period (from 1945 to 1972), and it approximately coincides with a peak in prices for lead, silver and zinc. Lead peaked at £140 per ton in 1954, and zinc at £122 per ton in the same year. Thereafter, prices tended to stabilise around the £100 per ton mark for lead, slightly higher for zinc; silver continued to rise.[63] By 1961, the town's population was still just above 31,000. The mining workforce was now below 5000. Despite major growth at the Broken Hill North and South Broken Hill mines, increasing mechanisation had capped the increase in the numbers of miners in Broken Hill.

By 1960, miners were working with mechanical loaders and tungsten-tipped wet drilling machines. The ore was transported via air-operated chutes onto battery-powered underground locomotives, then going through the crushers and mills, then being delivered to rail trucks via conveyor belt.[64] These 'concentrates' were then sent to Port Pirie for refining.

As the lead price moved from £22 per ton in 1948 to £35 in 1950,[65] the lead bonus for Broken Hill workers by 1951 was £18/17/6 per week – this was on top of the regular basic weekly wage of £10/9/6.[66] One of the markers of this period for Peter Bennett was the number of new homes constructed:

> [T]he price of ore went up and it was really going well. People were building homes as a result of it. A lot of the houses out at South Broken Hill here you'll see are a result of that money, and that income in the town generally.[67]

For Elizabeth Firth, newly married in 1956, Broken Hill's prosperity was further enhanced by housing loans provided by Zinc Corporation at 2 per cent. Even though she was expected to leave her employment (as a clerk at one of the major mining companies) upon marriage, her husband's position at a major bank and later at Zinc Corporation meant they had a comfortable life. Asked what it was like living in Broken Hill then, she replied:

> Well I think it was fairly good because of the working conditions and the wages that we received. Certainly helped; it helped us anyhow, and the fact that we didn't have children [meant] the lifestyle was very good. We did eventually have our little girl but we were married twelve years before we had her, and of course in that period of time we had pretty good holidays.[68]

Zinc Corporation was particularly active in local affairs. In 1947 the company began construction of 150 houses in the Railway Town area.[69] This was on top of the co-operative housing societies, which were established in 1937, and which also received company support.[70] By 1960 Zinc Corporation boasted that it had financed 'the building or purchase of nearly eleven hundred houses for employees'.[71] There was also a spate of company-built bowling greens and sporting amenities. Zinc Corporation also began offering subsidised annual holidays for employees and their families at Largs Bay, north of Adelaide, where the company had constructed a purpose-built camp. This activity (which mirrored BHAS's holiday camp at Weerona Bay for Port Pirie employees) further cemented the town's ties with South Australia.

Carole Michaels moved into a Zinc Corporation home in Broken Hill South in 1955: 'It's the last street – behind us was the regeneration area.' It was a steel-framed weatherboard home:

> My brother was doing his national service in 1955. And we moved into this lovely new house. Three bedrooms, still no running hot water, you still had to boil the copper and carry the buckets.[72]

Lorenzo Cester arrived in Broken Hill in 1956, aged thirteen. Like many postwar migrants, he had travelled by ship to Melbourne and then by plane to Broken Hill. He recalls the day he arrived:

> It was one of those typical Broken Hill days. It was a real red dust storm that day and my mother started crying. We were told we were coming to the bush. The word 'bush' over in Italy has got a different meaning to here![73]

The dust storms were still a part of local life, though the regeneration project from the late 1930s, led by Albert Morris and with assistance from the mining companies, slowly created a protective green belt that minimised their worst effects.[74] The 1961 Census records 441 Italian-born migrants in Broken Hill: 262 males and 179 females.[75] Cester's family moved into a shed for a few weeks, then to a house, sharing with another family. This was also a common experience for new arrivals.

Postwar growth and new migration policies had not modified local demography. While Mount Isa and the steel towns of Whyalla, Newcastle and Port Kembla became major destinations for migrant workers and their families, changes to the ethnic composition of Broken Hill were modest. In fact, there was a net decline in the numbers of European born in Broken Hill between 1947 and 1971. This is partly explained by the declining numbers of British-born migrants, but the percentage of European-born people at Broken Hill peaked at 6.4 per cent in 1954, reducing to 5.1 per cent by 1971. Broken Hill's demographic changes were at odds with the national trends over this period, during which the Australia-wide percentage of European-born people moved from 8.6 per cent in 1947 to 15 per cent in 1961.[76] Broken Hill's experience was similar to that of other towns where industries had flat labour demand and there were no major new industries and projects.

The housing stock that accommodated the workforce mostly dated from the turn of the century. Broken Hill homes were basic galvanised-iron structures, known locally as 'tinnies', lined with tin or hessian, or even old newspapers.[77] To one side there was a chimney of brick or iron: for warmth in winter and for cooking. Of the 6075 dwellings in Broken Hill in 1921, the most common materials for the outer walls were iron (2548) and wood (2172). Decorative pressed metal ceilings or wall linings were common, partly because they were light-weight, which meant lower transport costs. The roofs were overwhelmingly iron (5844), rarely wood (198). There were no tile or slate roofs, though in the suburbs of Sydney and Melbourne they were as common as iron.[78] The night cart system operated in Broken Hill with rear lanes or side lanes providing access. The degeneration of the topsoil and tree cover meant that sand and dust were constant companions. Some erected iron or even stone fences to stem the flow of the sand drifts. Moving sand plagued householders, and covered roads and rail lines. The major companies sent crews to work clearing accumulated sand from their fence lines.

By 1954 there were 8017 occupied dwellings in Broken Hill, with wood the most common outer wall (3172), followed by iron (2667), stone (936), brick (505), fibro (339) and concrete (264). We can assume that many of the brick, concrete and fibro-clad houses were more recent constructions.

May Watson's house, 45km out of town on a sheep station, was rather different. It was a three-bedroom brick house, with a wood stove, surrounded by large verandahs, and was built around 1940. It had two rainwater tanks for the house, and when she was a young girl the area around the house was fenced for a garden. This house was quite different from the house she lived in for four years as a single parent in the late 1960s: it was a stone house owned by her parents, and was divided into three flats. Stone-walled houses remain a distinctive – if minority – presence in town, one of the elements of the local landscape that reveal the influence of South Australian building styles.

Postwar politics

The BIC and the MMA dominated local industrial and political affairs. Their negotiating of regular three-year agreements outside the state arbitration system dated from 1925 and was a unique feature of industrial relations at Broken Hill. As author and diplomat Paul McGuire observed:

> The Hill is a curious town. There are two powers in it, the mining companies and the miners' unions ... I do not know any country in the world but Australia where the local nightman could be the chief citizen of no mean city, and without resigning his sanitary engagement. Sometimes I suspect that we are a democratic people, after all.[79]

This was a reference to Shorty O'Neil.

The fact that elements of Broken Hill's society and political economy were shaped by the BIC and the MMA was remarkable; it was unprecedented in Australian terms. BIC control of the labour market (favouring locally born men over non-local men and married women) was coupled with price control of items such as milk and beer. The preference clause for men, first instituted in 1931, continued into the postwar period, but was modified as demand for labour increased. 'A

Groupers' were men who had been born and had grown up in Broken Hill; 'B Groupers' were men who had married local girls and had been resident for more than eight years. Those who were not in either of these categories, or were otherwise unknown outsiders, were often spoken of as being from 'away', a term that is still used by older Broken Hill residents. 'Away' meant both those who had come to Broken Hill from elsewhere, and all places beyond Broken Hill.

Women were defined by their marital status. Single women could do paid work, but in some occupations, married women could not. This was common throughout cities and towns in Australia, but it took a particularly rigid and distinctive form in Broken Hill. The use of badge show days from the 1910s (where all union members had to wear a current union badge or they were unable to work), and the successful campaign to unionise the town workforce in the 1920s, meant that the BIC had virtually 100 per cent compliance. All paid workers were members of unions and all obeyed the rules, as set down by the BIC. As Ellem and Shields note, the public space for women contracted as a result of these rules, despite the efforts of second-wave feminists such as Labor party identity and Broken Hill's first female Alderman, Nydia Edes, who held office from 1962 to 1974 (after 1968 as a Labor Independent).[80] This social compact only began to unravel in the 1970s, under pressure from state and federal laws against discrimination; the locally negotiated three-year agreements came to end in 1986.

In the context of a town where young single men, occasionally outsiders, were recruited to work for lower wages or break strikes, the BIC's local preference rules embodied both an industrial commitment to a local fully unionised workforce, and an ideological commitment to settled family life with a male breadwinner. The BIC's movement into town politics and society was also the product of the suspicion of private enterprise resulting from the water and rail experiences. Without this context, later policies appear more ideological than practical. Ironically, in the postwar period the Broken Hill companies also supported suburban family life, in part by acquiescing to the BIC's local preference, but also through their own vigorous housing and community welfare programs.

The MMA and the BIC also increasingly dominated the civil society of Broken Hill. As we have seen, they involved themselves in housing, and set up joint union–company schemes in the health and accident fund area. They ran a dental clinic and a pension fund; the MMA also had its own sickness fund and the BIC acquired the *Barrier Daily Truth* in 1962.[81] From their high point in the 1930s, the friendly societies slowly lost out to the company–union combination, and to the increasing role of the state in social benefits and insurance.

Unions postwar

The identification of the town as a union bastion or the 'Gibraltar of unionism' continued into the postwar period, and if anything was strengthened. Allegiance to the union was both institutionally enforced and culturally ingrained. For mineworkers, it was difficult to imagine paid work without the union. The sense of isolation and the knowledge that the ownership, control and broad direction of the mining companies was organised from Melbourne, through Collins House and associated companies in Collins Street, added to a sense that the town was a collective of wage earners, well organised but ultimately set against powerful distant forces. The NSW government shared some of this identification as a distant, problematic outsider, and was still seen as less engaged than it should be. As a result, class and local loyalty were often expressed as complaints about the wealth that flowed out the region instead of being reinvested. These appeared in a variety of guises, from attempts to make the mining companies pay higher municipal rates to proposals for reform of mining royalties. As Peter Bennett expressed it: 'If Broken Hill built the nation as we called it, then the nation should start putting money back to rebuilding Broken Hill.'[82]

Class and local loyalties ran deep, yet the Broken Hill labour movement and its commitment to the rights of workers in others places also focused this class loyalty outwards, to other places and other industries. However, the commitment to internationalism had clear limits; the BIC had made its choice by closing its books to outside workers in 1931. In 1888, at the very beginning of local organisation, Broken Hill workers sent contributions to striking Newcastle coalminers, and in the following year to London dockworkers. When Mount Isa workers were locked out by MIM in 1964–65, Broken Hill workers committed £38,000 to the strike fund and senior labour movement figures went to the town to lend their support. Socially conservative and strongly institutionalised Broken Hill labour movement leaders such as Shorty O'Neil were suspicious of the charismatic union rebel Pat Mackie, but they instinctively sided with Mount Isa workers in their battle against a determined employer.[83] These class loyalties also encouraged unions to send money to worker movements in foreign countries. The local preference rules did not diminish these outward-looking class sympathies.

Postwar society

Just as there were two major political forces in town – the BIC and the MMA – there were also two moral worlds. There was a family-centred world with religion, friendly societies, cultural and sporting organisations. This was a sober, respectable society. The temperance movement was strong, drawing support from many of the non-conformist Protestant churches and the Catholic Church. In the 1947 referendum on closing hours, 7319 people voted for 11pm closing, but 6086 voted for 6pm closing. A similar referendum in 1954 actually gave a slight majority to 6pm closing (8632) over 10pm closing (7611).[84]

The first public art gallery in regional New South Wales was established in Broken Hill in 1904. The town has a justifiably famous artistic tradition, including artists J.C. Goodhart, Eric Minchin, John Pickup and Pro Hart.[85] This tradition built on older artistic traditions, such as the brass band movement, song and choral performance, and theatre. A Municipal Library opened in 1891, catering to a growing reading public and the self-improvement impulse that was strongly ingrained in union, church and community. That the council ran the library, eventually taking on the collections from the old Silverton Library and from the Mechanics Institute, was also indicative of a municipal socialism that sought to educate, and improve the life of the mind. The library was managed and greatly expanded from 1909 to 1913 under the leadership of R.S. Ross, a well-known socialist and journalist. In 1913, the library had 1461 registered borrowers, an average of 1000 volumes in circulation each week, and branch libraries in Railway Town and Broken Hill South. Fiction accounted for 60 per cent of all loans, followed by history, biography and travel (20 per cent), literature (10 per cent), sociology (5 per cent), and scientific works (5 per cent).[86]

Broken Hill has figured in a number of Australian and international films since the late 1960s, and the local council has recently established Film Broken Hill. Local newspapers, authors, and cultural organisations have created a lively literary scene. Many well-known journalists worked at the *Barrier Miner*, or the *Barrier Daily Truth*, or later ABC Radio, Broken Hill. Authors Roy Bridges, Arthur Upfield, Ion Idriess and Kenneth Cook all included the town in their works, and many had active links to the community.[87] Complementing these well-known names is a thriving local publishing scene.

The other moral world was the world of hotels, gambling and prostitution, peopled predominantly by young single men. Peter Bennett recalls:

> There were a couple of pubs on Argent Street where you know, you'd go past, on your bike or something at night and you'd hear the doors crash open and a couple of blokes would come out ... you know. A bit rough. But not in all pubs, only just the odd couple.

Bennett also remembers the gambling:

Everybody knows about it now. We used to have a two-up school, right, and it was good because when you went in there you were checked by the Cockatoo, who was the bloke sitting out the front to check you out, make sure you were sober or looked sober. You went in ... We used to go in and say I could have a quid, which was a pound note. You'd pay your 10 bob for a feed and you'd bet with ten bob, right? If you doubled your 10 bob you put it in your pocket and you've got your free meal and you've made a 10 bob profit, you know? Other blokes were really serious. They'd go in there and you'd see money all the way around them, and they'd have bets all the way around the circle. It was very well run.[88]

Known locally as 'the game', the two-up school was in the lane across Oxide Street, opposite the Astra Hotel. It continued until the state government took control and limited illegal gambling; 'the game' closed in 1984.[89]

Public health surveys from late in the 20th century show that the town then had high rates of alcoholism and domestic violence. One study of patients at the local psychiatric clinic found alcohol-related problems at a rate four times higher than the NSW average: some 13 per cent of men and 3 per cent of women had alcohol dependence or abuse issues.[90] More recent work has indicated high rates of domestic violence, low levels of trust in the community, and a general exclusion of women, youth and Indigenous people from leadership roles and community forums.[91]

This unruly, drunken, male-dominated side of Broken Hill can be overstated, and often appears to reflect a stereotypical view of mining towns and miners rather than reality. One *Sydney Morning Herald* journalist visited Broken Hill in 1953 in search of 'sin in the silver city', and found very little. The journalist noted that the rates of public drunkenness in the town were lower than in the rest of New South Wales: 'Drunkenness is not common, and it is doubtful whether the amount of gambling is relatively greater than in Sydney.' The journalist found no evidence of prostitution at all.[92] Although such activities are very difficult to follow in the historical record (a point which is relevant for all the towns in this book), newspaper records reveal only a scattering of court cases involving brothels or prostitution in the 1890s and 1900s. This broadly coincides with a period when large numbers of single miners lived in the town.[93]

The network of family, union, church and community was a broad-ranging support structure, usually available to those who needed it. Bennett recalls that those who arrived in the 1960s found a welcoming community and numerous points of entry. For Carole Michaels, whose father died when she was a child, in the mid-1950s, the community support was overwhelming:

> Dad died when I was nine, so probably that's the thing that makes me so loyal to Broken Hill. When dad died ... we received a lot of support. It was very sudden. He died without a will, and Mum had to find work.

For Michaels, family and church were the main forms of support:

> Once Dad died and Mum had to go to work there was two types of care that I had. I had wonderful assistance from all the family, like school holidays, [and] if I was sick ... And then the other side was the Nuns ... Mum would go to work at ten to eight and I would walk with her so I would be at school at 8 o'clock. The nuns would find all sorts of things for me to do. One job I used to have was cleaning the silver and brass in the church. And I love silver to this day and I still find it very therapeutic to sit there and clean it.

The close bonds of community and union power were a comforting embrace for many, but there were some who saw a darker side to it all, in particular to the enforced union solidarity. Bob Bottom grew up in Broken Hill, worked as a journalist for the *Barrier Daily Truth*, and then as a local ABC journalist in the 1960s. He clashed with the BIC over his reporting, and had a strong individual clash with BIC President

Shorty O'Neil. His book, *Behind the Barrier*, was a critique of collective authority flavoured by a sense of personal grievance. Railing against BIC control of the mines and the town, Bottom wrote that 'the state of affairs in Broken Hill demonstrates that even in a peaceful contented community, officials blessed with too much power can exploit blind allegiance and indifference to impose autocratic and sometimes despotic rule'.[94]

South African author Elspeth Huxley, who toured Australia in 1965, spoke with Broken Hill locals:

> It is impossible to winnow the hard truth from all you hear. Some of the instances I was given of trade union tyranny over private lives may have been influenced by personal prejudices, and by the national habit of grumbling at authority whoever exercises it.[95]

Others, however, noted that 'Broken Hill is the best town in Australia. People who tell you otherwise are traitors. Everyone is looked after.' Even Bottom grudgingly accepted that 'for all their faults, Broken Hill people have gained a well-deserved reputation for hospitality towards visitors and for generosity towards charitable causes'.[96]

Such comments agree with the oral history record, gathered many years later. Respondents evinced a general allegiance to the town, though they were not dismissive of, or unconcerned about, specific problems. The evidence gathered in the 1960s is expressed in language that is more emotive than the oral evidence, though: the reference to 'traitors' is the clearest expression of this. Perhaps it is a language of class solidarity that was more common in that decade than today.

The decline of mining

The mining boom at Broken Hill lasted from 1945 until the early 1970s. There were occasional downturns, but steady growth allowed good wages and employment prospects for young men in particular. The end came slowly. There were a number of high-profile layoffs between 1972 and 1981. Mining employment dropped from 4473 (including 157 females) in 1971, to 2062 (80 females) in 1986. The Broken Hill South mine ceased production in July 1972, with 650 employees losing their positions. Local business, union and political representatives pooled their resources, and the mine was reopened by them, trading as Minerals, Mining, Metallurgy, and on a much reduced basis, later in the year.[97]

The demography of the town changed as the mining industry employed fewer and fewer men. As mining declined, females caught up and then outnumbered males. At the 1986 Census, for the first time in 60 years, females slightly outnumbered males. The demographic profile of a booming mining town, with large numbers of single men and young families, had changed. In 1961 just under 7 per cent of the population were aged 65 and over; in 1991 it was just over 14 per cent. The workforce had aged and retired, fewer younger people were moving to the town, and more retirees were moving into town from the hinterland regions. Affordable housing and the availability of health services suggested a possible future for the town as a retirement destination, but a major battle over the state of the local medical services in 1992 indicated that the state government did not necessarily share this vision.

As the mines neared the end of their effective life, attention turned towards the past. Efforts to publish local history and preserve key heritage buildings began in the late 1960s. The Broken Hill Historical Society was founded in 1968, and foundation president Roy Kearns, a retired chief accountant from the North Broken Hill mine, published a three-volume history of the town. These moves coincided with a series of controversies over heritage-listed buildings, including the Town Hall. In 1977, the Delprat Shaft from the old BHP mine was reopened by local businessmen Ted Williams after a long struggle with state authorities to declare it safe for visitors to go underground. By 1986, the council had established a Heritage Committee and the long process of identifying, preserving and interpreting Broken Hill's rich heritage began in earnest. The past is now of considerable interest in terms of future economic development, but it is still alive in the consciousness of many locals too. One respondent was asked what history means to Broken Hill:

> It's its soul. I mean if people, even this generation, if they found out, knew how the place came into being and how hard people fought for it to become what it is, they would have a greater appreciation of the town and its ideas.[98]

Mining left a clear tangible and intangible heritage. There are other less positive legacies. Broken Hill is still dealing with the environmental legacies of mining. The regeneration project was an early example of environmental rehabilitation of the kind that is now common, but the ongoing exposure of the community to high levels of lead is a major challenge. While lead and other toxic substances are no longer released into the environment, the tailings and dumps, and the ubiquitous dust of The Hill, are not so easily expunged.

A vigorous society emerged on the Barrier. It was shaped by isolation, harsh working and living conditions and the cultures of union, friendly society and church. Town loyalty developed very quickly, initially expressed through the Progress Committee and appeals for infrastructure, but later almost principally through the organisations of the local working class. Locality fused with class in a most powerful way.

The compact between the BIC and the MMA in the postwar period shaped work, society and culture at Broken Hill, reducing the town's organisational diversity. From the 1970s, wider changes, including a reassertion of state control and the decline of the mines, conspired to slowly but surely fragment this settlement. Though class and union dominated the town's history until the 1970s, they should not obscure the religious, cultural and suburban complexities of its lived experience.

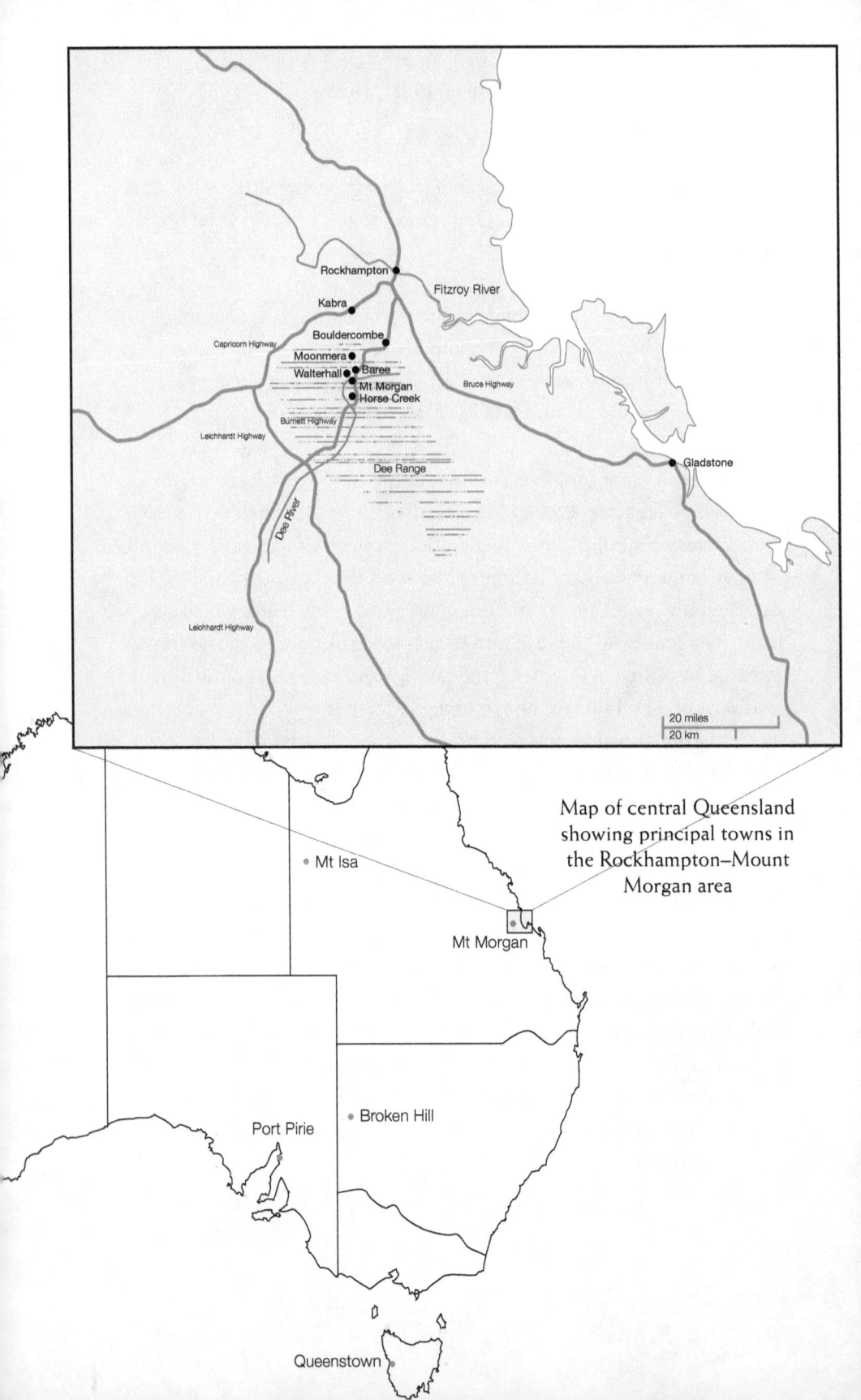

Map of central Queensland showing principal towns in the Rockhampton–Mount Morgan area

3

MOUNT MORGAN
In the thrall of modernity

It is early morning on a summer's day in December. Central Queensland steels itself for the heat that will follow. Up on the ranges a shimmering haze picks up from where it left off yesterday. The night was a short respite from the heat. Clouds roll in from the north, peering over the ranges that fringe Rockhampton, but they amount to nothing. The former mining town of Mount Morgan sits behind the highest part of the Dee Range, about 200 metres above sea level, nestled in a small valley amid the recovering bush, 39km southwest of Rockhampton. Brisbane is 500km to the south.

The mountain that gave rise to the gold, silver, and copper mining town of Mount Morgan was once some 387 metres (1270 feet) tall. It formed the western boundary of the town. Across the Dee River, on the side of the mountain, a town developed to service this mine. The average maximum temperature here is 27°C, with the peaks in December and January. The heat is moderated by the elevation, the mountains bringing cooling breezes at night; the average lowest temperature is 15°C with June, July and August being the colder months.[1]

All the contradictions of modernity are embodied in this place – the faith in progress and technology, together with periods of gloom and decline. After the initial excitement and wonder at the wealth of the mine, the trajectory of development was sporadic, punctuated by major downturns, halts to production, and disasters. The relative calm of the town today is misleading, and the view of one local historian in

1939 is instructive: 'Half a century's glamorous growth and tragic life, progress and retrogression; the rise and fall of a town [is] indissolubly linked with that of its only industry.'[2]

A local society and identity grew quickly here, in an effort to make this harsh world hospitable, but it co-existed with class, ethnic, fraternal and regional identities. In this amalgam of positions, localism was the dominant one, but exclusions based on race and class – even on suburb – kept the localist project incomplete. Modernity involved the creation and destruction of space, and such patterns were clearly evinced by this town.

Making a mine

In 1882 the Mount Morgan gold mine began production. Operations were overseen by a syndicate of six which included the three Morgan brothers and Rockhampton-based investors. In its early years the mine was primarily worked for its gold. The Mount Morgan Gold Mining Company was formed in October 1886 with capital of £1,000,000. The company had a tumultuous life: it added copper ore to its production in the early 20th century, and suffered disasters and labour disputes. It liquidated in 1927. After a five-year struggle, a new company, Mount Morgan Limited, under the guiding hand of the former managing director of the old company, Adam Boyd, restarted production with decidedly different labour management principles.

The early years were full of legal disputes over title and occasional tension among the original syndicate members. There were nine court cases which threatened the syndicate's title, high-profile events that were widely reported in colonial and English newspapers.[3] Around the fringes of the syndicate's land other companies were formed. Some were hoping to find payable ore from the same orebody; others were trading on the growing reputation of Mount Morgan in order to sell shares to an unsuspecting public. In 1886 the editor of the Rockhampton-based *Northern Argus* described the prospectus of the Mount Morgan (West) Gold Mine thus:

Mount Morgan

> Certainly no document has ever been more successfully penned in which fact and fiction have been perfectly blended ... there are evidences of a hand of [a] master in the art of deceit.[4]

Title disputes were common on mining fields, especially goldmining fields. They were part of the early development of most of the mining fields covered in this book, and only disappeared around the 1960s.

The company was registered as the 'Mount Morgan Gold Mining Company Limited' in October 1886. The Morgan brothers sold their shares to the others and after 1886 were no longer a part of the operation. Those who kept their shares – Thomas Skarrat Hall, a bank manager from Rockhampton, W.D. Darcy, a Rockhampton-based solicitor, and William Pattison, a local grazier – became some of the richest men in the colonies, as the mine returned fabulous profits. Early share trading saw prices of £1 to £2 per share, but by the end of 1888 shares peaked at over £16, and the company returned a dividend of over £317,000. Many of the original shareholders sold parcels of shares for huge profits and still had substantial holdings. These profits were channelled into grand homes in Rockhampton, Brisbane, Sydney, Melbourne and eventually London. Those who bought at the peak of the market watched as the share price dropped dramatically during the financial chaos of 1891 to 1893. This fall was helped along by the original shareholders selling large holdings.[5]

A stamping battery (a crushing plant using steam-driven piles to crush the raw ore) was erected in 1883 and the first successful treatment works to refine the crushed ore was established in 1886. The boilers to run the stamping batteries were heated by burning wood, so woodcutters were a common sight throughout the area. Easily accessible trees along the Dee Ranges were soon felled. The flow of the river slowly reduced as its water was used for the batteries and building dams became a company priority. The last and largest, the No. 7 Dam (known locally as the Big Dam), was built in 1900. The gold content was extremely high but the ore was difficult to treat and varied in composition. After a number of plants were trialled, the Lower Works was constructed in 1886 at the foot of the mountain.[6] Once milled and roasted, the ore was mixed with chloride of lime and sulphuric acid to separate the gold. This process gave off chlorine gas. Ore shoots that ran from near the top of the mountain were replaced in 1888 by an aerial ropeway which carried the ore to the Lower Works. An Upper Works (another chlorination plant) was built in 1888, further up the mountain and thus closer to the ore supply. The larger West Works, another version of a chlorination plant, was constructed in 1896, as was the Mundic Works. The Lower and Upper Works were replaced by 1902. By 1901 the mountain was surrounded by plant and equipment on the company's holding.[7]

At the turn of the century concerns about the declining yields of gold forced a reconsideration of the future of the mine. Company officials were slow to recognise the importance of copper. After the allure and status of gold, the shift to copper seemed like a step backwards. However, richer copper-bearing ore was found in 1902 and drilling in 1903 proved the existence of high-grade copper ore.[8]

Blister copper (that is, 98.5 to 99.5 per cent pure copper) was first produced in 1902, and commercially viable copper ore from 1904. The company finalised plans for a new copper smelter at Mount Morgan, and also began planning an electrolytic refinery to treat its blister copper. In 1907 the company, in co-operation with the German firm Aaron Hirsch und Sohn, decided to construct a refinery. The site chosen was Port Kembla, on the south coast of New South Wales. Mount Morgan supplied blister copper to Port Kembla from 1909 until 1927. A number of Mount Morgan families relocated to Port Kembla, again indicating the strong links between these mining and industrial towns – financial, trade-related and social.[9]

By 1908 Mount Morgan was the most productive gold mine *and* the most productive copper mine in Queensland, with investment tentacles reaching out to other states and beyond.[10] That year 2470 men were employed at the mine and the new copper smelter, and 120 men were engaged in cutting timber for stopes in the mine, firewood, and occasionally furnace fuel (coal had substantially replaced wood in

1898 with the arrival of the railway).[11] The copper smelter required a new stack to emit fumes from the blast furnaces. The stack was completed in 1905 and stood at a height of 76 metres (250 feet). It became a significant symbol of the town and remains today, somewhat reduced in height owing to safety concerns. The town of Mount Morgan developed – mostly on the eastern side of the river – to cater for this large and diverse industrial workplace.

The region

The region had been open to white occupation from the 1850s, and available for selection from the 1860s. There were a number of goldrushes in the region, starting in the late 1850s. There was another goldrush at Crocodile Creek in the late 1860s, closer to Ironstone Mountain (today's Mount Morgan). This area is now the small town of Bouldercombe, 12km from Mount Morgan. Many miners soon drifted away, though, as richer finds were uncovered at Charters Towers and on the Palmer River.[12] There was a freehold property on the Dee River owned by John Coutts Gordon. It seems Gordon never lived there, and it was abandoned by the mid-1870s. There followed a complicated story of family intrigue, claim and counter-claim, the end result of which was the Morgan brothers gaining possession of the mountain, which they renamed in their own honour.[13]

All of this hardly amounted to extensive exploitation of the area: there were periods of rapid growth followed by retreat or decline. If there were not general economic problems, or government policy problems, there were droughts and floods. This only heightened the delight that followed news of a major gold find in 1882. As one 1897 survey noted: 'a town thrives at the foot of the mount, and the whole district has been regenerated by the enormous expenditure of the Company'.[14] The reference to 'regeneration' is significant. Mount Morgan's development was a sign of a revival in an area where the rural economy had not prospered.

The areas that did develop through pastoralism, gold and

coalmining were a great boon to Rockhampton, which became a major economic hub, the centre of pastoralism, meat processing, financial services, maritime and railway trade. By 1860 Rockhampton had been proclaimed a municipality, and by 1891 the population had grown to 13,380. The town continued this steady growth in the ensuing decades.[15] It would play a key role in the development of Mount Morgan: Rockhampton supplied the capital and commercial skills, and Mount Morgan supplied the raw material for Rockhampton's wealth.

The comparative isolation of Mount Morgan provided space for economic and civic development beyond the mine and its smelters, especially as getting to Rockhampton by road was difficult before 1898: the track between the base of the mountain and the top of the Dee Range was known as the 'Razorback' and had a bad reputation. Various proposals for a rail link to Rockhampton were initially unsuccessful. They were not helped by the difficult financial circumstances of the 1890s, but they were also opposed by Mount Morgan retail interests. A transport link into Rockhampton would threaten their trade.

In 1898 the Queensland Railway Department constructed the Kabra to Mount Morgan rail line. The years before the construction of the line allowed local commercial enterprises to establish themselves. Even after 1898, the difficulty of the rail journey gave local organisations some advantage.

The wealth of the region drove ambitions for a new state, and a Central Queensland Separation League was formed in 1889, supported by well-to-do pastoralists at Rockhampton, including successful Mount Morgan investors. For these men, the wealth of Mount Morgan was an argument for the separation of Central Queensland from the rest of Queensland.[16] Many regional politicians were Rockhampton-based supporters of separation. After 1886, most of them were also shareholders in the Mount Morgan company. A.J. Callan was the member for Fitzroy from 1889 to 1902, and a company director. John Ferguson, former Rockhampton councillor and mayor, was the member for Rockhampton from 1881 to 1888, later moving to the Legislative Council and the federal Senate. He was another shareholder who made fabulous profits from the mine.[17]

This drive for political autonomy in Central Queensland meant that the region was especially prone to place-based politics. Rockhampton would be a competitor to an emerging Mount Morgan identity. From the perspective of Mount Morgan, regional autonomy was so closely aligned with regional wealth and company ownership that it was difficult, even impossible, to separate the two.

Making a town

Development around the mountain was sporadic and unplanned. Wood cutters camped along the Dee River and houses were scattered along the creeks and ridges on the foothills of the mountain. Some clustered along the river for access to water, and others were close to the mine works to allow easy access for workers.[18] In 1884, Monckton's Hotel was operating at Mount Morgan, and like many of the early hotels, it was an important venue for local civic organisations. Even the churches used Moncktons: the Catholics held their first mass at Mount Morgan, in 1884, before constructing their own church in 1887. The Primitive Methodists built a small church in 1885 and later claimed to be 'the first to hold religious services, and to build a church at Mount Morgan'.[19] Service for the Presbyterians dated from 1887, and the church built in 1890. The Anglicans built their church in 1889, with support from the mining company and its senior staff, including general manager James Wesley Hall, brother of syndicate members and major shareholders Thomas Skarrat Hall and Walter Russell Hall. By 1889 the Salvation Army and the Baptist church were also active.[20] Next to Moncktons, the Queensland National Bank established a branch in June 1887, no doubt influenced by Thomas Skarrat Hall, the manager of the Rockhampton branch of that bank. By 1886 there was a sizeable population at 'South Calliungal'. The township name was formally changed to 'Mount Morgan' in 1889.

There was rapid growth from 1886 to the early 1890s. The Census of May 1886 estimated a population of only 632, but by the following year well-placed observers such as the *Morning Bulletin*'s Mount Morgan correspondent suggested a population 'approaching 2,000 souls'.[21] The same year general manager Wesley Hall reported that 500 men and 300 contractors were working for the company. Numerous new buildings were constructed: hotels, halls, churches, a hospital and commercial premises. The company built three substantial residences for staff in 1889. That year the *Morning Bulletin* noted: 'The rapid forward strides the township has been making of late are really worthy of note.'[22] According to the 1891 Census, the town's population had increased to 3514 – the Census probably underestimated the floating, itinerant populations which were common in growing mining towns.[23]

The shacks and temporary dwellings scattered through Tipperary Point and Red Hill, and the hotels and boarding houses gathered at the mine entrance, were in great contrast to the company residences, which were built on the mine side, above the river and overlooking the township and the mine works.[24] In 1927 a child of one of the company directors who visited Mount Morgan recalled the general manager's house: 'The manager's residence was a big rambling house of weatherboard with verandahs everywhere built on the slope of a hill overlooking the mine.' The residence included a croquet lawn, two chimneys, and electric lights.[25]

Despite the local power and status of senior staff, their overall influence was dwarfed by that of the Melbourne-based directors. This higher level of power and influence was exemplified in Mount Morgan by 'Carlton House'. Carlton House was built in the early 1890s, and used until 1927. It was a large nine-room timber Queenslander with wrap-around verandahs finished in iron lace work. It had two chimneys and an impressive garden, even trumping the General Manager's residence. It was the venue for staff gatherings and for entertaining visiting officials, directors and dignitaries. Unlike other company houses, Carlton House was sited in the heart of the developing town. It was set on a natural rise, on elevated piers. There was no clearer indication of company dominance.[26]

Mount Morgan

The noise of the stamping batteries, fumes from the works (initially chlorine gas and later sulphur dioxide from the copper smelter) were an ever-present feature of local life. Mine sirens and hooters marked out the rhythms of the day. The mine ran day and night, in three eight-hour shifts, during good times:

> Mount Morgan is a veritable hive of industry. From dawn to dark, and from nightfall to sunrise, the work never ceases, and the ponderous clang of the machinery smashing up the golden stone is most annoying to visitors at first, but they very soon get used to it ...[27]

The regular fortnightly pay days also shaped the commercial and town-based economy as flush workers and their families headed to the stores to buy supplies. The overwhelming physical presence of the operation, plus the widespread belief in its success and wealth, meant that later halts in production, and then liquidation, had a profound effect, which still resonates in the town.

By 1908 the town centre, in and around Morgan, West and East streets, had emerged from a number of smaller suburban sites. It had the services and facilities of any sizeable town. Hotels and boarding houses were particularly important for itinerant labourers, and catered for all classes. Boarding houses were one of the few areas in the commercial sector where women played a central role. While women were possibly present as staff in the town's 22 hotels, they were primarily male spaces. Apart from the stores which supplied daily necessities such as bread, vegetables and groceries, Mount Morgan also had a number of commercial and industrial establishments which manufactured or prepared food, including a slaughterhouse and aerated water firms. In 1911 these probably accounted for many of the workers who listed their occupations as 'manufacturing' and 'undefined industrial'.

The town centre, which was dominated by wooden buildings, was ravaged by fire many times. This was a common experience in 1880s mining towns, and occurred at Broken Hill and Queenstown as well as

Mount Morgan. The School of Arts, for example, burnt down twice. The original 1889 building burnt down in 1896, and its replacement burnt in 1923. It was replaced in 1924 by a building that still stands. The fire of 1896 wiped out many businesses in the East Street area, as did the 1923 fire, which destroyed a general store, a drapery and an auctioneering business, among others.[28]

Taking root

The speed with which workers and their families established themselves in housing of all kinds was matched by the speed and energy put into building the organisational elements of town society. As we have seen, the churches were early arrivals. Fraternal organisations also played a particularly important role. The role of the Catholic benefit society, the Hibernians, is a case in point. The society was active at Mount Morgan from 1888, with the ubiquitous hotelier, William Monckton, a member. Some 80 people attended the first meeting in the Catholic church. Apart from its medical and insurance benefit functions, the Hibernians also organised St Patrick's day activities, including a parade and sports. An annual ball was inaugurated in 1888, and was a popular social event, attracting some 80 couples in 1906. By 1908 the Hibernians had their own hall, and 188 members, including a ladies' branch.[29] Part of the town was known as 'Little Ireland', including the shacks in Tipperary Flat, Tipperary Point and Red Hill. The region had a higher than average concentration of Catholic worshippers. By 1921, 39 per cent of the local population professed adherence to the Catholic faith, well above the Queensland average of 24 per cent.[30] Organisations like the Hibernians provided fellowship, financial support, and social activity for transferring and new members. The churches, sporting groups and other fraternal organisations had a similar impact. It was a ready-made community, perfect for a new town.

Politics and the company

A central division ran through most mining and industrial towns, and changed little through the 20th century: the gap between 'staff' and 'wages' employees. Staff were usually professional, technical men or senior and middle managers. They were often paid on monthly or, at the top of their echelons, yearly contracts, and roughly equated with the middle class. Wage employees were usually hired on a short-term basis, and worked while there was work to be done. They were general labourers as well as men with trades or specific skills that gave them a premium on top of a labourer's wage.

In this early period the Mount Morgan company staff were the leading force in all aspects of local society. These men were engineers, accountants, assayers and shift bosses. Workplace and political networks also fused socially in the small town. For example, the first mayor of the new municipality (in 1890) was none other than James Wesley Hall, Mount Morgan general manager and brother of Thomas Skarrat Hall and Walter Russell Hall. The second mayor (in 1891) was Francis Bunny, assayer and chemical analyst with the company, and brother of Australian artist Rupert Bunny. Francis Bunny was educated in Melbourne. He was the son of a Victorian judge, Brice Frederick Bunny, who was also a former politician and local alderman. Bunny was a Captain in the defence force volunteers at Mount Morgan, and was married to the daughter of Roger Lisle, the mine manager. The staff families had a tight social network, and marriage within the staff ranks was common. Bunny moved to Western Australia to work on the Boulder goldfields, and died in a shipping accident *en route* to Melbourne on the steamer *Orient*.[31] Most staff were Anglicans, and many were also Freemasons.

These men had local status as well as formal management positions, and brought the wages employees and their families into a kind of paternalist framework, where workers felt obliged – or compelled – to profess their respect. A clear example of this occurred following the marriage of Wesley Hall in 1889. On return from his honeymoon both the company and the town offered an almost ritualistic public expression of his fine qualities and the loyalty of the workers to his wise leadership. The town band played outside his home, and a crowd of thousands cheered and applauded the appearance of Wesley Hall on the balcony to thank them for the welcome, and the band for their music.[32]

'Loyalty' is probably the best word to describe the social relationship between the company officers and the employees. In this period the company was the focus of allegiance and identity. The interests of the masters and those of the men were seen to be in sweet harmony. Capital and labour were reconciled ... or at least labour had reconciled with capital. Historian Betty Cosgrove even found evidence of a great pride among early residents in the wealth and prestige of the company, expressed through a vicarious attachment to the clearest symbol of company standing, the magnificent Carlton House.[33]

But there was another side to the apparently free expressions of loyalty, as revealed by the visit of Josiah Thomas, a prominent union official from Broken Hill. Thomas toured Central Queensland in 1892 looking to raise money for workers affected by the 1892 dispute on the Barrier. At a meeting in Rockhampton it was noted that:

> on account of there being no organised unions at Mount Morgan, much terrorism is displayed by employers and their agents under the freedom of contract that obtains. The men at Mount Morgan have as yet done nothing, but Mr. Thomas and the local men have no doubt that, when things are started on Wednesday, the Barrier men can safely anticipate great financial assistance from there.[34]

Mount Morgan men were reportedly afraid to attend Thomas' public meeting, and even the conservative *Morning Bulletin* commented that:

> 'Freedom of contract' was so utterly abused there that such of the men as desired to hear Mr. Thomas met in fear and trembling in the open, under cover of darkness, just as men sneak away to

witness a prize fight. That gives, we believe, a fair indication of the conditions under which the Mount Morgan men have been living ...[35]

It was not until the late 1890s that this company hegemony was seriously questioned by Labor candidates for municipal and colonial elections. As the number of workers grew, and organisation among them increased, a class identity emerged to challenge company dominance. Since 1889, much had happened in the Australian colonies that encouraged workers to see their interests as separate from those of their employers. This change led to the growth of Labor parties and new unions.[36] At the 1899 Queensland elections, member for Fitzroy (and Mount Morgan director) Alfred Callan was challenged by T. Wightman, a local Labor candidate who was nominated by the Democratic League. Callan won the poll, with votes at Mount Morgan being 653 to 428. Wightman had done well, despite support for Callan from the company, and senior staff advising the wages employees how to vote.[37]

Callan lost the subsequent election in 1902 to Henri Cowap, another Labor candidate. Cowap was a Mount Morgan local, having moved there in 1888. Labor introduced a new language of class to Mount Morgan. For example, speaking at a public meeting during the 1902 election campaign at the Mount Morgan School of Arts, Cowap advocated taxation reform:

> The millionaires could draw their money from Mount Morgan,
> live at home in peace and comfort, and enjoy all the luxuries that
> the world could bring them; but all they paid to the Government
> was five per cent. Should not an absentee tax be bought about to
> lighten the burden of taxation on the people?

This concern over the disappearance of Mount Morgan wealth persisted throughout the 20th century. The grand homes built with Mount Morgan money in Rockhampton, Brisbane and London left a bitter taste for those in the ramshackle huts along the Dee River.

Few directors and shareholders were active locally: Walter Hall was an exception. The nationally famous Walter and Eliza Hall Trust, funded by the sale of 350,000 Mount Morgan shares from Hall's estate, spent the majority of its funds in New South Wales and Victoria, with Rockhampton and Mount Morgan receiving what one historian calls a 'mere pittance'.[38]

In this period the language of racial exclusion was particularly strong. There were at least twelve Chinese market gardens and two Chinese fruit stores at Mount Morgan, but the new Labor representatives were as likely to criticise the Chinese as to criticise the actions of the Mount Morgan company. There was a history of anti-Chinese sentiment in the region, with some violent attacks on the Crocodile goldfields. This was given added impetus by the deportation of Pacific Island labourers through the *Pacific Island Labourers Act* of 1901. The link with earlier goldrushes and the mining of gold at Mount Morgan, as well as the geographical proximity of the goldfields, lent these racial sentiments a sharper edge. At the School of Arts speech, Cowap spoke disparagingly of the local Chinese. His comments are worth quoting at length, as are the crowd responses:

> There were a few Chinese in Mount Morgan, but not quite so many as in other places. (A voice: "Good job, too.") But the first thing that a stranger landing in Mount Morgan from the train saw was a Chinese garden, or rather two Chinese gardens – on either side of the road. And the first store he saw in coming to the main street was a Chinese store. Such things should not be allowed in a country like this. (Applause.) Every man in Mount Morgan could help to put away the Chinese. If the working men would not deal with them he did not think very many of the swells would.[39]

Cowap referred to the 'White Queensland' policy approvingly. There were clear racial boundaries around the working class that Labor sought to represent.[40]

Indigenous history

There were other exclusions too. A company history of Mount Morgan written in the late 1930s begins with this statement: 'we must think of white settlement advancing like an incoming tide, wave after wave; each wave overlapping a little further than the last'. This was a telling description of colonial expansion. Pastoralism, then selection farming, and finally large-scale mining each embodied a more intense and encompassing form of land use. Small groups of Bayali, the local Indigenous people, had found some spaces at the edges of the colonial occupation in the late 19th century, but with the discovery and development of Mount Morgan, each square foot of land was pegged, mapped, fought over in court, and finally mined.

The area was known to Indigenous people as Bundoona. The Bayali people of the Dee Range and Dee River have close links with the Gangulu, from the Fitzroy River and Dawson River areas. In the 1880s and 1890s, small groups of people from both groups still camped by the Dee River, despite the town's rapid growth, but it was soon polluted by mine and township runoff. Some worked for white settlers as domestic help or as stockmen. The Morgan family employed Aboriginal labour at Mount Morgan and General Manager Adam Boyd employed Aboriginal women at the mine manager's residence.[41]

By the 1920s, there were Aboriginal camps near the Mount Morgan slaughteryard, at Horse Creek, Box Flat and near the racecourse. Aboriginal society was as dynamic as settler society. Some of the interviewees who spoke to local researchers in the early 1990s referred to marriage between white, Chinese, Malay and Indigenous people. Many lived in families linked to several cultural traditions.[42] There were large numbers of Indigenous residents; this is still the case in the early 21st century.

Progress unleashed at the 'Golden Mount'

The pace of change was rapid, and the mine's and the town's growth had regional and national effects. Mount Morgan was the richest gold and copper mine in Queensland, and was regularly celebrated for this. It pre-dates Broken Hill, and in a few years Mount Morgan financiers would invest in Broken Hill and Broken Hill wealth would in turn buy Mount Morgan shares. Mount Morgan became the favourite topic of travellers, journalists and atlas editors in the late 1880s and 1890s.[43] In fact, in a climate of colonial and regional boosterism, which was particularly strong in Queensland, it was a common boast that it was one of the finest sights in Queensland. Hume Nisbet, a Scottish-born artist and popular novelist who travelled Australia in the 1880s and 1890s, wrote that 'one of the most wonderful sights about this part is Mount Morgan, which is a mountain of gold'.[44]

The exploitation of the mine's resources, and the men and machinery that made it possible, unleashed great change.[45] In 1897 the company employed some 1100 men and production continued 24 hours a day. Mount Morgan was well known in investment and banking circles in Brisbane, Sydney, Melbourne and London. Recent research indicates that it was the 15th largest company in Australia in 1910, with assets of almost £1.4 million.[47] This is important in terms of the impact of the later decline and failure of the company: that can only be understood if one appreciates the ubiquity of the celebration of the 'Golden Mount'.

The Great War and the 1920s

By 1914 prospects for the Mount Morgan company were looking solid. The earlier plant had been replaced by more modern and efficient blast furnaces. The works had been reorganised and redesigned. The company had secured supplies of coal from the nearby Central Queensland coalfields. The Queensland government extended the rail line from Mount Morgan through to the Dawson Valley to access the coal fields in that area.[48] But the war created significant upheaval for Australian copper producers. German firms, including the company's partner in the Port Kembla refinery venture, Aaron Hirsch und Sohn, were active in the Australian and world copper market. As Chapter 1 outlined, Australian and British capital replaced German interests with support from federal and state governments by 1915. The Mount Morgan company was involved again at Port Kembla, this time supporting the formation of Metal Manufactures Pty Ltd in 1916; it would use Australian-produced copper for much-needed rod and wire products.[49]

In terms of financial conditions, once the German contracts were dealt with, the war brought increased demand for copper. This benefited both the Mount Lyell and Mount Morgan companies. Copper prices peaked on the London Metal Exchange at £160 per ton in 1916. These prices did not always translate into actual contract prices, but the constant demand at least meant stability. On the gold front, the Commonwealth government controlled the market and prohibited export. Also, many men from Mount Morgan enlisted, which created a labour shortage.[50]

The end of the war bought further economic problems. The British government ended guaranteed minimum prices for copper in December 1918. The copper price fell to £76.5 per ton on the London Metal Exchange in 1921, and languished at the £60 to £70 until 1928. These poor prices shadowed the company's every move for the next decade. Grand plans for redesign and redevelopment all needed capital, which these prices could not support. At the same time, management had to reduce costs. But by the 1920s, the men at Mount Morgan were not keen to compromise. The years of company heavy-handedness, the perceived opulence of staff conditions, and the Carlton House garden parties, left a legacy that would poison later labour relations.

By the end of the war, the cost of living had increased, but wages had not even kept pace with inflation. Workers had become more radicalised – especially those in mining, transport and construction. At Mount Morgan the Labor cause, established in the early 1900s, was strengthened by municipal success and by continued industrial and community organisation. Fatal mining accidents in September 1908 (seven men killed) and November 1908 (five killed) had given further impetus to union activity. The Australian Workers' Union (AWU) appears to have taken over from the General Workers' Union in 1913, and the union secured preference in employment from the Queensland Arbitration Court in 1917. Between 1894 and 1917 the AWU had transformed itself from a union of shearers and other pastoral workers to a general rural workers' union.[51] Amalgamations with smaller mining industry unions had helped this expansion. The local branch was an active one, as were the Electricians' Union and the Carpenters' Union branches. By the 1920s a Labor league was well established, and Labour Day processions and an associated sports carnival rivalled other days in the social calendar for popularity. The interests of the wage earners were well represented, and class identity was clearly expressed through these local organisations.

The numerical dominance of the wage earners was the basis of this class identity. By 1921, one in three males at Mount Morgan (or 1159) were engaged in 'quarrying and mining'. There were 150 in manufacturing industries and 106 in 'undefined industrial'. For females, 148 were domestics, 71 were in manufacturing industries, and 11 were in mining and quarrying. This compares to relatively small numbers in staff and commercial occupations. There were 90 males in professional occupations, and 204 in commercial occupations. For females 99 were professionals, and 68 were in commercial occupations.[52] Many of the town's storekeepers were closely aligned to the majority working class either through social and familial networks or through the Labor League. The most high-profile example was J.H. Lundager, a commercial photographer and newspaper editor who was active in the town's Labor League from the 1890s and was mayor of Mount Morgan in 1908.[53]

Given the cost pressures on the company, problematic labour relations, and the growing disenchantment of the men with arbitration and compromise, the industrial disputes of the 1920s were no surprise. Throughout the country industrial disputes broke out – on the wharves, in the mines, and in the shipping industry.[54] In December 1920 the company applied to have all industrial awards – covering 1200 employees – suspended, and also sought a 20 per cent reduction in pay.[55] The Mount Lyell company in Queenstown had also sought a wage reduction, and there is evidence that Mount Morgan kept a close eye on developments there. By October 1921 the Queensland Arbitration Court had approved the wage reduction at Mount Morgan but the local men, supported by Jimmy Stopford (MLA and Minister for Mines), disagreed with the principle that a business could be enabled to start up by offering below award wages.

By 1922 there was a major downturn across the country. Mines were idle at Broken Hill, Queenstown and Mount Morgan, and furnaces were shut down in Port Pirie and Port Kembla. Unions had initially opposed the wage reduction, but two company employees convened a meeting – with the aid of the mayor, T.T. Cornes – which censured the union officials and declared that the mines should reopen, on terms favourable to the company.[56] Ultimately, it was government intervention, in the form of an offer to drastically reduce freight rates, that saw the mine finally reopened in March 1922, after almost a whole year shut down. The government also agreed to subsidise the wages of employees by rebating some of the rail freight earnings back to the company. So when the men returned to work the minimum wage was £3/5/- a week, plus an additional 12 shillings per week covered by the government rebate.

The future for the company still looked bleak. *The Argus* in Melbourne supported the wage reduction and criticised 'militant unions', but still warned that:

> Unless the company is particularly successful in effecting economies, and bringing about the saving anticipated by using coal from its own Baralaba mine, it would appear that work is to be resumed at a substantial loss as long as copper continues at the present low level, even after making allowance for the £1,100 weekly rebate that is to be allowed by the State Government.[57]

The mine had suffered from the long closure – regular maintenance had fallen behind, experienced miners had left the town, and the ore-body groaned and moved unpredictably in the deep shafts. A plan to convert to open-cut methods and change the smelting process was developed in 1924, but it required £1,000,000 in investment.

The labour dispute of 1925 and the fire which followed soon after were complex and controversial events. The directors asked for a further 20 per cent reduction in wages – in fact, they had decided to close the mine. Workers saw this as an attempt to blackmail the Industrial Commission. Angry meetings and pickets, plus a confrontation between senior staff and up to a thousand workers, ensued. Smoke was seen rising from the workings on 12 September. The mine was on fire. Deep in the mine's shaft, at the 574 foot level, the fire raged. There was little that could be done. Strikers volunteered to assist but the men could not access the fire. The fumes and heat were too much. So the mine was flooded. Two months later, in early November, the company began pumping out the millions of gallons of water that had finally overcome the fire. The mine was a wreck: machinery was burnt or destroyed by water, and the timbering was weakened and unpredictable.[58]

These events were interpreted in many ways. For years, men at the mine blamed General Manager Adam Boyd for starting the fire. Colin Herberlein said:

> Many believed that Adam Boyd had set fire to the mine because as a mining engineer and general manager he realised that the economic viability of the mine in the long term was the open cut method.[59]

A battle over the public memory of the dispute and fire saw the miners putting out stories of management duplicity, and the 'official' published sources either glossing over the events or laying the blame at the feet of the strikers.[60]

In 1927 the company went into liquidation: the shareholders and the early directors had gouged too much profit out of the mine, and left its equipment out of date. Also, there had been no concerted drilling campaign to establish ore reserves, and smelting techniques for low-grade ore were not always successful. Labour management strategies were rudimentary compared with the programs at Mount Isa and Port Pirie, both of which will be covered in detail later. Most other Australian mining and industrial firms managed to struggle through far worse conditions during the height of the Depression.

The challenges of the 1920s

The modernist dream of progress and improvement is always shadowed by the nightmare possibility of decline: perhaps faith in growth will not be enough. Now the nightmare was upon Mount Morgan. The labour dispute of the early 1920s was the first sign of trouble. The men had returned to work on lower wages, but the reprieve was short-lived. There was a general feeling of decline and downturn after the 1925 fire and the 1927 liquidation, as long-term residents left, businesses closed or relocated, the mine closed, and the Depression began. As one oral history respondent, Arthur Timms, noted of 1925: 'That was the start of the depression in Mount Morgan.'[61] In July 1929 the *Morning Bulletin*'s Mount Morgan correspondent noted the impact of the decline:

> Of late the town has been very depressed, worse perhaps than any other period of the close down. The withdrawal of banking institutions and business houses, besides the departure of citizens of many years' residence have all in their turn cast the hopes of a revival down to zero, and added to the impression that Mount Morgan was just another worked out mining field.[62]

Bill Butcher, interviewed in the early 1990s, recalled a large number of house fires in this period, as desperate householders tried to collect on insurance:

> [W]e would get out on our bikes at night waiting for the fires to start ... We would race to the scene of the fire – no fire brigade in those days – and watch the houses burn.[63]

Given the powerful messages about progress, wealth and development since the mine's opening, the decline had a strong emotional impact. Charlie Sheen, interviewed by local researchers in 1992, remembered feeling the twin effects of the closure of the old company and the arrival of the Great Depression. A former apprentice electrician at Mount Morgan mine, he had to leave town and finish his apprenticeship elsewhere. Relatives wrote to him to keep him in touch with events at Mount Morgan:

> I was out the bush one day, getting wood for the wood stove ... I sat down and thought what is to become of me ... there is no work ... there are no lights in the town ... even the gas we had ... the fellas who went around and lit the lamps on the lampposts, Herman Heinkel, council foreman, no lamps ... it was dark.[64]

The memory of the gas lights going out and the men who had lit them losing their jobs is a distinctive rendering of this dark time in the town's history. The lights were going out all over Mount Morgan. These same lights had stood for production and employment on the nearby mountainside for more than 40 years. Only 25 men were employed at the mine between 1927 and 1932. Houses prices declined sharply and those who were still employed could purchase a home cheaply. Others took up land on miners' homestead leases, which were very cheap and could be converted to freehold after 30 years. By June 1933 the town's population had declined to 3262, from a high of 10,000 in 1911.

The decline was felt keenly because Mount Morgan had been the centre of a vigorous social and civic world, full of garden parties, football teams, a vast array of civic organisations, and annual picnic events and regular dances at the School of Arts, one of the town's many halls, or in nearby settlements such as Bouldercombe and Baree.[65] Moreover, many locals had witnessed the rapid growth of the 1880s and the 1900s. The dramatic drop in population was a shocking experience, which the challenges of the Depression only made more difficult. In this context the man who rescued the mine would become a local hero.

Remaking the mine – the 'new' company

A newly listed company, Mount Morgan Limited, under the leadership of the previous general manager, Adam Boyd, purchased the mine for £70,000 in 1929, and reopened it in 1932. The original company was now known universally in Mount Morgan as the 'old company' and Boyd's Mount Morgan Ltd was the 'new company'. The new company adopted a number of different strategies, including the committee-style system for workplace consultation, new facilities for workers (including drinking fountains), and improving local sporting and recreational opportunities. Mindful of the past, the new welfarist labour management strategy, borrowing elements from Port Pirie, Queenstown and Mount Isa, was formally announced in 1934, and was to be funded by 7.5 per cent of all distributed company profits. This supported the construction of a recreation ground (including a swimming pool) at Mount Morgan (later known as 'Welfare Park'), life insurance policies for all workers, a dental clinic, cash bonuses at Christmas time, and a Works Picnic on Boxing Day at Emu Park, on the seaside east of Rockhampton. The works picnic was later moved to Good Friday. The company also funded the upgrade and maintenance of other local facilities such as a bowling club, the cricket pitch and tennis courts. According to the Rockhampton-based *Evening News*, in the old days there 'was neither confidence nor trust on either side', so the welfare policy was a welcome sign of a new approach.[66]

These policies tried to extend the kinds of facilities and services offered to staff to all workers, yet in common with other companies from Queenstown to Port Pirie, the distinction between the two remained and was in other ways reinforced. At Mount Morgan the Staff Club Rooms still served as a place for staff to relax, enjoy billiards and read, for instance, and so was still somehow separate from town society. But the welfare policy, with its emphasis on unity and consultation, set the tone for labour management at Mount Morgan for the next 40 years.

This largesse generated a significant amount of goodwill towards the company. Many recall that it was the company which supported the town, in particular through the efforts of Glenister Sheil, general manager from 1950 to 1964. Yet the policy was not simply about benevolence for the sake of it; it was seen in terms of the costs and benefits. E.T. 'Taffy' Badham, an official for the Engine Drivers' Union in the 1940s, was supportive of the scheme but noted: 'it was used against the unions, admittedly ... every time they made an application for a variation of the award, the Company pointed out how they had the scheme operating'.[67]

The new company stayed afloat, but growth was modest, and new mining methods were of course less labour intensive. The population stabilised at around 4000 in the postwar period. By 1954 there were 4152 people: 2168 males and 1984 females. There were 882 males and 30 females who listed occupations in the 'mining and quarrying' category. The total 'mining and quarrying' workforce was 912, a reduction from the 1159 recorded at the 1921 Census. There were 63 males in transport, since Mount Morgan was a key railway site with at least nine gangs of a dozen men working in the 1950s.[68] There were 80 males and 81 females in 'commerce', and 29 males and 54 females in 'amusements, Hotels, etc'.[69]

There were few new houses built because the population decline meant that there were plenty available. Houses had been removed from the town in the late 1920s and new construction appeared to have just replaced these losses, with total dwellings stable at around the 1000 mark. In 1954 there were 1054 occupied and 65 unoccupied dwellings. The overwhelming majority of homes were constructed primarily of wood (830), with fewer made of iron (124) and fibro-cement (92). There was still one tent or canvass dwelling in 1954.[70] Houses were on tank water unless they were lucky enough to find water in their backyard and sink wells. The mine continued to dominate the local economy: in 1959, Mount Morgan Ltd still employed 82 per cent of the shire's male workforce and owned 50 to 60 houses.[71]

Working at the mine post-1945

From 1945 there was a sustained period of prosperity that produced regular employment and good wages. Copper prices rose strongly after 1945 with a peak of £347 per ton in 1955. There was a significant drop, to a low of £211 per ton in 1958, but from the mid-1960s prices went to well over £400.[72] Gold prices were more closely controlled by governments, with all Australian-produced gold sold to the Australian mint at approximately £15/12/6 per fine ounce. After 1951 limited amounts of gold could be sold overseas, but a rapid decline in gold prices in 1954 meant that only small amounts went to that market.[73] The Commonwealth government responded with the *Gold Mining Industry Assistance Act 1954* (Cth), which gave assistance to miners whose output was more than 50 per cent gold. The *Copper Bounties Act 1958* (Cth) gave further assistance to Australian mining companies by levying a tariff on imported refined copper at particular price points. Both Acts were indicative of the extent of government support for the mining industry, and had significant effects in mining towns such as Mount Morgan. Prosperity was both hard won and strongly supported by Commonwealth and state governments.[74]

There were three principal types of workplaces at Mount Morgan Ltd. The open cut had been a part of the Mount Morgan operation since 1882. It was a more significant part of the new company's operation and replaced underground mining altogether by 1965. Sydney ('Syd') Wall was working as a truck driver in the open cut in 1940, earning £14 per fortnight. He soon moved to drilling work, placing and setting explosive charges. Wall remembers many of the men killed or injured in open-cut operations, from rock falls, equipment failure and faulty explosives. (Wall's extraordinary memory of the men who died or were injured is confirmed by newspaper searches.) There was a strong pressure to continue mining, despite the dangerous conditions and practices. Wall recalls a bonus system that rewarded shift bosses for moving more ore than other shifts.[75] Diesel trucks now moved the vast majority of ore. Fewer miners were actually mining.

Between 1894 and 1909, 26 miners were killed at the mine. One oral history respondent suggests that 130 men died at the mine, a figure which appears broadly correct but to which we need to add the many more who were injured or maimed. A former nurse at the Mount Morgan hospital, Lorna Moulds, recalls the men in 'D' ward in the late 1930s and 1940s:

> They took a lot of nursing. These were not only underground men but those from the smelters as well. The disease is not called phthisis now; it is referred to as a type of asthma, but was referred to as silicosis then.[76]

Underground mining continued to 1965. The orebody went as far down as 335 metres (1100 feet), and miners recalled the mine 'speaking' to them with cracking and groaning timbers, and small rock falls. Thick choking dust was common after an explosive charge or 'shot'. As in most large mines, rats, which Charlie Shannon remembered as being 'as big as cats', lived in the shafts near the crib rooms, which is why miners' cribs were made of metal.[77] If the town was plagued by fire, the mine was plagued by water. The Mount Morgan mine was especially susceptible. Steady rain would produce a runoff into the shafts that was hard to shift, slowed production, and added to costs.[78]

In the two mills and in the copper smelter (which opened in 1939), men did not have the travails of underground labour, but there were other challenges. Despite diverting most of the furnace gas to the bag house, the smelter gave off sulphur dioxide which could hang about the workplace and the town in the right weather conditions. The heat of the furnaces when they were being charged or tapped was often oppressive, particularly in summer. The mills worked seven days a week, three shifts a day, producing copper concentrate for the smelter.

The noise of the crushers, the washing section and the grinders in the mills could be deafening. Mill superintendent Bryce Hargreaves recalled 150 to 160 workers in the mill:

> Most of them were local lads, trained operators, [and] because of the age of the town, 50 per cent or more were related. It was not surprising to find the father on the shift as the shift foreman, a son on the same shift as well as his brother. That was how the town worked.[79]

There was a strong demand for young men to work at the company. Jim Leigh started at Mount Morgan as a sample boy in 1945, earning 12/6 per week. All nineteen boys in his class at the Mount Morgan State School worked at the mine after they left school. In the 1960s Mount Morgan was a major provider of apprenticeships, with most of the 50 to 60 places being filled by locals. Young women worked as typists and switchboard operators between leaving school and marriage. There were 12 to 14 at the company; other women worked in the town's retail sector.[80] The mine had a very good reputation as a place to learn a trade, for apprentices and for young engineering employees, who were encouraged to do further training. This was a tangible contribution to the local society and the wider economy.[81]

Making a town – post-1945

The sounding of the works hooter marked out the days and nights with precision. Many locals could recount the hooter times with unerring accuracy: three blasts at 6.30am, two at 7.30am, and one at 8am to announce the start of day shift; smoko at 9.50am; lunch at 12 noon for 30 minutes; day workers finished lunch at 1pm; 4 pm was the end of day shift; two more blasts at 7pm for a 7.30pm start for the evening shift; two blasts at 11pm and one at 12 midnight to indicate the start of the night shift. For many, shift work became a way of life. The sounding of the hooter became such an established feature of town life that its loss in 1981 was widely mourned. The hooter now has a prominent place in the local museum.

These years are remembered as prosperous, good times. Donny McDonald started at the mine on the road gang after the war, following many years on a farm or doing rural labour:

> Yes, very productive years, about 1500 men working there in the fifties. There was a job for everyone. I saw the open cut go deeper and deeper, right to the depth it reached before it closed.[82]

Patrick O'Regan, who started a family newsagency and electrical contracting business in 1956 recalled: 'Business was pretty good then. When I went to work at Mount Morgan [1951] there were 1100 men here and in 1956 there was a lot of overtime being worked.'[83] Shift work generated extra income that flowed into the town stores. The water supply remained a problem. A new water supply system, the Fletcher's creek scheme, was completed in 1954, but later dry spells proved that a further expansion was necessary.

A steady rate of growth plus flat or declining levels of employment meant that there was not the labour demand of towns like Mount Isa, Newcastle and Port Kembla. As a consequence, the ethnic makeup of Mount Morgan remained strongly Anglo-Celtic. In 1966, out of a population of 3869, only 4 per cent (or 158) were born in Europe. This is the lowest rate among the six towns under scrutiny in this book. That same year there were only 13 born in Asia (including 6 in China).[84]

So stories of arrival are more muted at Mount Morgan. The great period of growth and population development was in the 30 years after 1882. After that there was no major increase in the scale or scope of operations of the kind that Mount Isa experienced. Small numbers of migrants, principally from the United Kingdom and Ireland, arrived in the 1950s and 1960s. There were 114 born in the United Kingdom and Ireland at the 1966 Census.[85]

Allegiance to the town comes across as a quiet pride, but the tumultuous history and the fact that the Mount Morgan mine has now ceased production complicates it, making expressions of loyalty less straightforward.

That allegiance was further complicated by the smaller suburbs and satellite settlements around the edges of Mount Morgan township, including Struck Oil, Baree and Walter Hall. Within the town boundaries, distinct areas also emerged, such as Tipperary Point and Red Hill. Loyalties to these smaller geographical areas were revealed by Arthur Timms (later Sir Arthur), who grew up in Struck Oil, 10km from Mount Morgan. Timms recalls travelling to Mount Morgan on Saturday evening and seeing the Struck Oil boys delighting in scattering the crowds by riding down the centre of James Street. He recalls rivalry between the Struck Oil, Red Hill and Baree 'mobs'. 'At the same time,' Timms noted, at the big football matches 'if you barracked for Rockhampton and you were a Mount Morgan person, one of the women would hit you with an umbrella. So you learned to barrack for Mount Morgan.'[86]

Timms' recollection is indicative of the way that these smaller allegiances could be put aside when it came to competition with larger regional rivals. George Dial recalls that in the late 1940s, 1000 people would travel down to Rockhampton to see Mount Morgan play the Rockhampton rugby league team.[87] Mount Morgan had a strong rugby union tradition, but from the 1920s rugby league became more important. Dial's recollection highlights the fact that if anything could fire up Mount Morgan locals it was competition with Rockhampton.

Decline and de-industrialisation

In 1960 the chairman of Mount Morgan noted: 'Everyone knows, a mine is a wasting asset.'[88] Even during more hopeful periods there was an understanding that the life of the mine was limited. In 1934, general manager Adam Boyd noted at the works picnic that 'there should be work for the next twenty years if not longer'.[89] This put limits on the rhetoric of progress and growth. Uncertainty also encouraged others to think about alternative futures for Mount Morgan. While the shift from industrial/mining activities to tourism is clearly evident from the

1980s and 1990s, as early as 1935 the Minister for Mines and former local AWU representative and MLA for the area, Jimmy Stopford, spoke about the town's potential appeal to tourists:

> I have just spent two delightful nights in Mount Morgan, and I don't understand why Rockhampton people and business men in particular do not realise the value of the climatic conditions of Mount Morgan.[90]

Uncertainty also generated unusual political coalitions. The main target for lobbying was the Queensland government, which was such a key supporter of new mining and industrial projects that major state-sponsored infrastructure such as ports and railways often accompanied private sector investment. Local union and company representatives often found themselves lobbying together – for a better water supply, lower freight rates, or a new rail link. As early as 1908, J.H. Lundager, as mayor of Mount Morgan and long-time Labor party member, led a deputation urging the Minister for Railways to construct a railway line linking Mount Morgan with the coal deposit to the southwest, at Baralaba and Blackwater. The then general manager, G.A. Richards, undertook to purchase a set amount of coal from the state coalmine at a set price, which sealed the deal.[91] In September 1930 the Mount Morgan Co-operative Mining Committee approached the deputy Prime Minister, F.M. Forde, seeking federal government assistance to help the new company begin production. These broad-based efforts were crucial in the re-opening of the mine in 1932 – the federal government had approved loans of up to £15,000.[92]

These things also had social and cultural effects. Patrick O'Regan, a migrant from Northern Ireland who arrived in 1951 and started as an electrician at Mount Morgan, found that some houses were painted and neat while others were left without much care or attention:

> We came up in the train of course, but as I had a look around it, a lot of the houses were in need of repair. A lot of the houses were in

need of paint. At the same time, a lot of the houses that I went into were quite nice inside. I'm not saying that they lived in squalor or anything like that. When I asked them why didn't they do their houses up, they said, oh well, the mine could close anytime.[93]

Planning was difficult. Many lived with the somewhat overconfident assumption that the mine would go on forever. Oral history respondents, especially staff who had access to geological information, noted that some workers were floored by the announcement of job losses in the 1970s and 1980s. Others were preoccupied with the mine's future even during the boom years, finding it difficult to settle and fully commit to life in one place. It was an understandable dilemma, especially as stories of the 1920s were told and retold. They seemed to give credence to a belief in the fallibility of the mine and its management: someday fate would deal another blow to the mine, just as it had done in 1925.

1968 onwards

This period was characterised by changes in mine ownership, and the slow decline in workforce numbers at the mine. Peko-Wallsend Ltd took over Mount Morgan Ltd in 1968. Peko-Wallsend was a diversified holding company with interests in mining, processing and transport. For the first time in the mine's history, Mount Morgan was just one part of a broader network of operations. After 1974 there was a scaling back of operations as world copper prices trended downwards: they peaked at US$2053 per tonne in 1974 but fell to $1232 per tonne in 1975, remaining around that level until a slight recovery in 1979.[94] Mining ceased at Mount Morgan in 1981. The smelter continued treating local tailings and ore from Mount Chalmers, and Tennant Creek in the Northern Territory, until 1984, and a separate tailings treatment project continued. In 1988 North Broken Hill Ltd acquired Peko-Wallsend and operations ceased at Mount Morgan in November 1990.

On the wider stage, the copper industry in particular faced challenges. New metals such as aluminium and nickel, and new industries such as the oil and gas sector, were attracting more investors and increasing amounts of state and federal government support. Aluminium was a major competitor to copper, and new smelters and refineries backed by strong state government support opened in New South Wales, Queensland, Victoria and Tasmania from the 1950s. Australian miners and smelters also faced increasing competition from Japanese refineries, which were able to undercut Australian refineries. From the 1960s, more and more Australian copper ore and blister copper was bound for Japanese refineries.

The coincidence of the layoffs at Mount Morgan with the growth of the Central Queensland coalfields meant that some affected workers moved to Blackwater and Emerald, and to the port of Gladstone. About 50 men lost their jobs from the 'Black Friday' closure of No. 2 mill, and all up 200 were retrenched in 1976. As Bryce Hargreaves, the former mill superintendent noted: 'Mount Morgan more or less staffed the coalfields.'[95] Another 150 were retrenched when the smelter closed in 1984.[96] This all led to a steady population decline, from 4000 in 1961 to 2974 in 1981.[97]

It was not just declining labour demand which reduced Mount Morgan's population. In the postwar period, as roads improved and car ownership spread, it became possible to commute to Mount Morgan. There were already a number of men driving in or catching buses from nearby settlements, but there were also some commuting from Rockhampton. George Beattie came to work at Mount Morgan in 1966. A Scottish migrant from Glasgow, he settled in Rockhampton and regularly caught the bus to Mount Morgan. His overwhelming impression was that 'In Mount Morgan, the people were related to nearly everybody else.'[98]

In 1976 Mount Morgan Ltd announced that closure was imminent. A project to continue to treat tailings was seen as a 'reprieve'.[99] By 1981, production in the open cut had ceased, and in 1990 the final owners of the mine site, Carter Holt Harvey, concluded treatment of tailings. Thereafter, the focus turned towards rehabilitation, conservation and planning for a future without the mine.

Mount Morgan

For those who had enjoyed the prosperous years after 1945, the changes in the 1970s were hard to accept. There seemed to be a weakening of the social fabric of the town, and an influx of 'newcomers'. This brought with it a very specific lament about the end of old Mount Morgan. The town had had multiple influxes of migrants, but only as the old order began to decay did the term 'newcomer' emerge. Trevor Stock, who was born in Mount Morgan and whose father and grandfather had worked for the old and then the new company, expressed this quite specifically. He noted how:

> in the earlier days you could walk up Morgan Street, the main street on a Saturday morning and it would take you all the morning to get through Morgan Street by the time you had a yarn with this one and that one ...

As retrenchments started and new residents moved to buy up cheap housing, this old social formation appeared to be under threat:

> but now there are so many newcomers in town you can walk through Morgan Street in three or four minutes.[100]

Nonetheless, many of the newcomers were warmly welcomed. Richard White arrived in 1966 and lived at the staff quarters, on the western side of the river: 'It was a quaint little town, and people were very friendly.' This was a common sentiment expressed by oral history respondents.

After the decline in mining, housing became cheap at Mount Morgan. This attracted home buyers who were less well off, including the elderly and Indigenous people. Indigenous Australians earned less, on average, and had unemployment rates up to three times higher than non-Indigenous people, so a move to Mount Morgan was often a chance to buy a home. A number of oral history respondents interviewed in the early 1990s believed that large numbers of Indigenous people moved to Mount Morgan in the 1960s and 1970s, working for the Mount Morgan Shire Council or (less often) for the mine. Census figures confirm this. By 2006, 11.7 per cent of the town's 3198 residents identified as Aboriginal or Torres Strait Islander.[101] Cultural and family links to Rockhampton, and west towards the Dawson Valley, Biloela, and Woorabinda continue.[102]

Sometimes, modernity's almost religious faith in progress can no longer be sustained. A new narrative can appear: the 'flawed rise'. This account focuses on some fatal flaw – in the case of the mine, that it was always plagued by industrial and other trouble. Charles Copeman, a former managing director of major mining companies, was born and educated in Brisbane. His mother grew up in Mount Morgan, and he recalled his mother's impression of the place:

> Her impression of mining was that either Mount Morgan was on fire, flooded, on strike, or closed for some other reason.[103]

These are memories of Copeman's mother reinterpreted through his own concerns and perspectives. This emphasis is certainly different from many local memories, and analysis of the company's post-1945 record shows there were no major strikes. Copeman's sense of the mine's difficulties helps him explain the fate of the mine: in the end, the mine and its workforce brought closure on themselves. This argument is seen in numerous company publications which blame the workers for the 1925 fire and the 1927 liquidation.

Mount Morgan lives with this complex legacy: it is a mix of rapid growth and sharp decline; of grand plans and dashed hopes; of great wealth and poverty. The period from 1945 through to the late 1960s was an exception – these were years of stability, good wages, and regular employment. Being from Mount Morgan, no matter whether one was a wages employee or on staff, and no matter what organisations one was a member of, produced town allegiance. Likewise, though Indigenous identity was a complex mix of regional, cultural, and family loyalties, being from Mount Morgan was part of it.

The town still stands. It has outlasted the original economic reason for its existence. Residents and the state government now grapple with the environmental and social legacies of 100 years of mining. In 2008 the Shires of Mount Morgan, Fitzroy and Livingstone were amalgamated into the larger Rockhampton Regional Council, a final victory for the large regional centre which has, for so long, shaped the development of Mount Morgan. The mountain is now a deep lake, leaching concentrations of heavy metals from the open cut. The Dee River, now but a trickle, is slowly recovering. While industry destroyed the mountain, it created a town, which has been left more or less intact. It evokes the past in a powerful way, through its tangible remains and its atmosphere. With appropiate support and funding, heritage tourism may well become a new part of the town's economy.

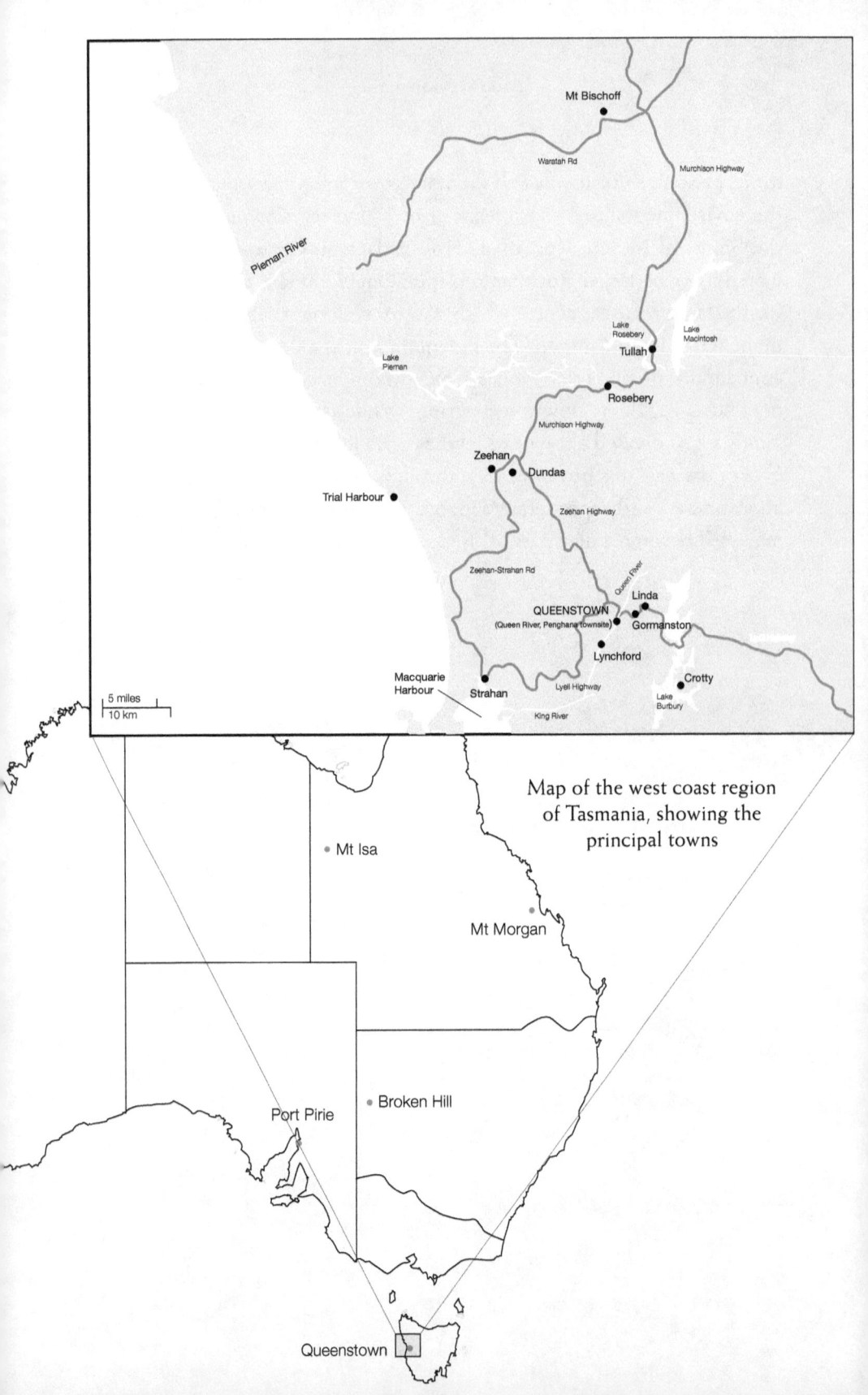

4
QUEENSTOWN
'They've got to come here and they've got to learn about it'

In the wet temperate hills near the west coast of Tasmania, a squall comes up the Queen River Valley, bringing chilly wind-blown rain in late summer. The mining town of Queenstown is in this valley; it has been here for more than 100 years. Restored miners' cottages, with heritage colours and bed and breakfast signs, sit next to roughly constructed kit homes with corrugated iron carports. There is a relatively new tourist facility, in the form of a $30 million heritage train, the grand old hotels, and of course, the mountains, where the majesty of nature has been topped by the power of human industry, at least until recent re-growth reclaimed the mountainside. The rain eases to a gentle patter, trickling on tin roofs. A pleasing gurgle of running water drains towards the Queen River, turning it a darker shade of orange. The wind dies and a patch of dazzling blue appears above the summit of Mount Owen.

Historical work on Tasmania concentrates on the 19th century, covering Aboriginal/settler relations, and the unique convict system that dominated white settlement from 1803 until the 1870s. Many of the island state's finest historians, including James Boyce, Kay Daniels, Hamish Maxwell-Stewart and Henry Reynolds, have made national and international reputations for their work in these fields.[1] Tasmania in the 20th century, especially its labour and social history, has attracted far fewer scholars.[2] A history of the Mount Lyell mine and its operations was the first published monograph by a young Geoffrey Blainey in 1954. That book, the first in a long line of commissioned company histories, probed the machinations of the Mount Lyell company, but it had less to offer on the workers and the community they built. Labour historians such as Charlie Fox contested Blainey's view of labour relations at Mount Lyell, and later historians provided new perspectives on mining growth and development.[3]

The community of Queenstown remains poorly defined, but locals are aware of state and nation-wide perceptions. As one respondent noted: 'Tasmanians have got these preconceived ideas about the West Coast.' When asked to elaborate, he said, 'Well we've got two heads around here. We're stupid. We're rednecks.'[4] Another respondent spoke of the outside perceptions of Queenstown:

> I know people think we're you know, hillbillies, but like I say to them well, do you love the town you live in? Cause I do ... I try and tell them, don't come here and condemn it ... But I don't think it's right ... they've got to come here and they've got to learn about it. And find out why the hills are bare.[5]

Beginnings of an industry

The 'Iron Blow' was discovered and pegged in 1883 by three prospectors: J.S. Karlson, and William and Michael McDonough. The blow was a rocky outcrop on one of the spurs of Mount Lyell, and would eventually yield rich gold, silver and copper-bearing ore. The three men were prospecting for gold and had found small amounts in a creek below the outcrop. That year there had been a small goldrush to the Lynchford area, 10km south of the future site of Queenstown, but no substantial deposit was found.[6] By 1886 there were numerous gold prospects being worked on the road from the port town of Strahan to Mount Lyell, including Lynchford, the King and Queen Rivers, and Linda Valley (later renamed the Mount Lyell Goldfields). Launceston financiers and gold seekers were already heavily involved, as they would continue to be.[7]

Mining at the Iron Blow and other nearby claims had its share of misfortunes and lucky breaks. The early syndicates were unstable, short of cash, often deeply factionalised and focused solely on gold. Mount Lyell, like Mount Morgan before it, saw its fair share of claims and counter claims, some of which played out in the Waratah Goldfield Commissioners Court or in Hobart.[8] After early success, a gold-stamping battery was laboriously transported from Strahan through Lynchford to the base of Mount Lyell, and then over the mountain to the Iron Blow. The conditions were difficult, the rough tracks from the coast at times impassable. However, the returns from the battery were poor.[9]

By the early 1890s, Broken Hill-based investors, including Bowes Kelly and William Orr, had become interested. And it was copper, not gold, that attracted their attention. After various changes in legal structure, a Melbourne-based company, the Mount Lyell Mining and Railway Company Ltd, was registered in 1893.[10] An extensive find of silver ore bankrolled the company and assured its early viability. The reduction works opened in 1896, and a railway line from Strahan to the works was completed in 1897. The company absorbed its main rival, the North Lyell Mining Company, in 1903. By 1910 Mount Lyell was the ninth largest company in Australia, with assets of over £2 million.[11] Other serious rivals, such as Comstock and Prince Lyell, had sold out to Mount Lyell by the early 1930s. By then, the Mount Lyell company dominated the town of Queenstown as the single major employer.

The west coast

On the west coast of Tasmania a rich mining tradition and a number of mining and port locations created a unique and complex urban environment. It is an isolated area with a stark and occasionally harsh environment. Mild summers give way to cold and wet winters: the average annual rainfall is around 2404mm per year, ten times more than the Broken Hill average. At Queenstown, the temperature rarely goes above 30°C, and winter snow storms can close the roads and reduce overnight temperatures to below 0°C.[12] The coastal areas are not as wet or as cold. Hobart, on the more densely settled east coast, is 270km away on the Lyell Highway, which was only completed in 1932. Launceston is 250km to the north, along an equally winding road – the Murchison Highway, which opened in 1963 – that passes through spectacular country fringing the high plateaus of the Cradle Mountain area.

Gold sparked interest in the region. Small finds were uncovered at the King River, the Pieman and Queen Rivers, and at Lynch's Creek (later Lynchford). A forestry and milling industry, using Huon pine and King William pine as well as eucalyptus, also attracted interest.[13] The isolation and the harsh conditions made prospecting and forestry work difficult, and deterred all but the most determined. Tin was discovered at Mount Bischoff in 1871 and the town of Waratah developed there. By 1891, Waratah had a population of 1420.[14] To the south, the silver-lead mining town of Zeehan, founded in 1882, was given a significant boost by the 'silver mania' that flowed from the rich Broken Hill find in 1883.[15] Government investment in a railway connection to Strahan, completed in 1890, solved the ever-present transport problem, and by 1891 Zeehan boasted a population of 1965. The town grew rapidly in the following decade, reaching over 10,000 in 1901. Zeehan became the hub for a number of smaller mining settlements, such as Dundas, which was connected by narrow gauge rail to the large centre. Here, as elsewhere in the region, railways dominated the transport networks. Dundas was 5km east of Zeehan, and had a population of 1080 by 1891. The ore deposit there was soon worked out, though, plagued by high costs and a significant underground water problem. Zeehan too began a swift decline just before the Great War.[16]

To the north of Queenstown, settlements based on gold, lead, silver and zinc, such as Rosebery and Tullah, developed from the late 1890s. These mines benefited from the Emu Bay Railway, a privately owned line which ran from Zeehan to Burnie, a port town in northwest

Tasmania. The section from Burnie to Mount Bischoff was completed in 1878; it was later extended south to Zeehan.

By the 1890s the region extended from Burnie in the north through to Strahan, in Macquarie Harbour, on the middle of the west coast.[17] Further south and inland were the pristine wilderness areas which became part of the Franklin-Gordon Wild Rivers National Park and the Southwest National Park in the late 20th century. The rugged terrain was covered with thick wet rainforest or open country scrub and had poor soil, so there were few rural occupations except timbergetting and fishing. The majority of the population lived in the mining and port towns. Unlike Broken Hill, there was no hinterland as such. At the 1933 Census, only 2 workers from Queenstown listed their occupation as farming.

Strahan, Trial Harbour and Emu Bay (Burnie) all developed infrastructure to service the isolated mines. This was often replicated, though, as companies, syndicates and towns competed for trade and economic development. This is also true of Queenstown: a number of small towns developed near the orebody and there were two major railway lines, owned by competing companies.

Mount Lyell and nearby Mount Owen were developed as separate fields by two major companies, often pitched in ferocious competition. The North Lyell Mining Company, formed in 1897 under general manager James Crotty, worked the Gormanston and Linda Valley area, west of Queenstown. In 1900, the North Lyell company completed a railway from the Linda Valley to the Kelly Basin in Macquarie Harbour. Crotty was an Irish-born prospector who had been heavily involved in the earlier development of the Iron Blow. He was determined to develop the mining field. Blainey called him 'an eloquent, excitable man-about-town'. He was also a minority shareholder in the Mount Lyell company and soon retired to Melbourne, like so many of the early investors.[18] The town that developed to service the North Lyell mine was named Crotty. It was a small settlement hewn out of the forest, and was also the site of the North Lyell smelters. After the amalgamation of the North Lyell company with the Mount Lyell Mining and Railway Company in 1903, Crotty was abandoned; eventually the site was inundated by the construction of Lake Burbury and the Crotty dam hydroelectric scheme, which was completed in 1993.

After 1903, Gormanston continued as an outlier of the Mount Lyell's Queenstown-based operations, with miners from there having easier access to the shafts until the incline was tunnelled from the western (Queenstown) side of the mountain to the eastern (Gormanston) side in 1928. Gormanston had a preponderance of miners, some of them single men, and was always a working-class settlement; few of the staff lived there. After 1928, Gormanston slowly declined, with the population dropping from 1589 in 1921 to 341 in 1986.

On the eastern side of Mount Lyell, the Mount Lyell Mining and Railway Company had constructed its own railway to Strahan. This narrow-gauge line was steep in parts, and used the 'abt' or rack railway system, the same system as the Mount Morgan line. Robert Sticht, an engineer from Illinois, was appointed by the new company and arrived in 1895, with his new wife Marion Sticht (née Staige). He pursued a form of smelting that used the sulphur and the pyrites in the ore to charge the furnace. Ideally, the pyritic furnace did not need additional charge (wood or coke). This was a great benefit, as the costs of transporting fuel were high. Even wood was not always easily accessible because of the rough and sodden terrain. Sticht's new smelters were ignited in late June 1896.[19] The smelters helped develop a national and international reputation for the company even though they eventually required the use of some coking fuel.[20] To provide this, the Mount Lyell company invested in a coke furnace at Port Kembla in New South Wales in 1899.

The Port Kembla cokeworks was one of many investments in plant operations beyond the Queenstown area. The company invested throughout the west coast of Tasmania and in other states. In 1905, the company opened a fertiliser plant at Yarraville, west of Melbourne. From 1915, head office was in Collins Street, Melbourne. Despite the Melbourne head office and these scattered ventures, Queenstown remained the company's main industrial base. In 1898 Sticht took up residence in 'Penghana', an impressive Federation-style home – it now operates as a luxury bed and breakfast and restaurant, owned by the National Trust.

Making a town

Queenstown was surveyed by the government in 1895, and work on a road from Strahan to the new town site continued until the end of that year, when new owners of the subdivided blocks at Queenstown began building their properties. Clearing the valley for construction timber and for furnace fuel proceeded rapidly in the 1890s.[21] One substantial building, Percy Waxman's hall and restaurant, was in existence from at least September 1895, providing the venue for the first local social events and meetings.

Queenstown then received an unexpected boost from a tragedy. Further up the valley on the Queen River, the town of Penghana had developed, at the site of the Mount Lyell reduction works. In November 1896, Penghana, home to around 1000 people, burned. One man was killed and all except three of the wooden buildings were destroyed. Both the government and the Mount Lyell company had already decided to close Penghana to new settlement in order 'to prevent', as one visiting journalist noted, 'litigation from persons who would have taken up ground here, and might be poisoned by the fumes of the smelters'.[22] The fire hastened these plans. After the fire, Penghana residents moved to Queenstown, many with aid from a colony-wide fundraising effort.[23]

After the Penghana fire, development proceeded rapidly in Queenstown. In early 1897, a 'Progress Committee' represented the needs of the town to the Minister of Lands during his visit.[24] Later that year, a 'Town Board' was organising a roll of rateable properties and planning urgent sanitation and road works.[25] The local correspondent for the Hobart-based *Mercury* noted that 'shops are being erected in Orr Street for hairdressers, fancy goods dealers, and others, at a rate which will before long compel later comers to look to other streets for building sites'.[26] By 1903 Queenstown had a population of 5293. It was the third largest town in Tasmania behind Hobart (34,917) and Launceston (21,606) and just ahead of Zeehan (5252). There was a commercial centre and there were 14 hotels.[27]

The churches were active in establishing congregations and building churches and church halls from 1897. They helped establish the town's social and community life. In December 1899 Robert Sticht laid the foundation stone for St Andrew's Presbyterian Church. St Andrew's, like many other churches, struggled to attract long-serving ministers. Even in 1956 St Andrew's had not kept a minister for longer than two years. According to a church publication, the '[r]igours of climate had not favoured long ministries', and in other cases 'ministers wished their children to have a higher education than that available here'.[28]

The Anglican church, St Martin's, was closely associated with the senior staff at Mount Lyell. St Martin's was opened in late 1898, with the Bishop of Tasmania presiding over the consecration; it was where the children of the ever-present Robert Sticht were baptised.[29] By 1921, 49 per cent of the total population professed the Anglican faith.[30]

The Catholics established their congregation in early 1898 with the opening of St Joseph's, though services (and baptisms) were conducted in a temporary building from at least 1897. As at Mount Morgan, the Hibernian Friendly Society was also active. The Catholic community had the benefit of a long-serving priest, Father Michael O'Regan, in the early years. Father O'Regan presided over growth and consolidation, with the construction of a new presbytery, a convent, and schools at Queenstown and Gormanston. By 1910, there were approximately 1700 people of the Catholic faith at Queenstown and Father O'Regan reported that 110 received the sacraments once a month, and 1410 came to church at least once a year. Many of the miners were Catholics, as were several prominent small businessmen, including Irish-born John Moroney, who ran the Family Hotel from 1898, and local chemist John Joseph Ryan.[31] By 1921, Catholics made up 26 per cent of the local population.[32]

A Hospital or Medical Union was established in 1897, receiving subscriptions from the Mount Lyell Company (£200), from company director and investor Bowes Kelly (£100), and from local fundraising efforts. With the Tasmanian government agreeing to match locally raised money, the Union found itself able to construct a cottage hospital with six beds. However, the expansion of the town, and in particular the need for facilities to treat injured miners, meant that this modest facility was soon overextended. In a circular appealing for funds from Mount Lyell shareholders, the secretary of the Medical Union wrote that 'the available accommodation has been seriously overtaxed, and tents have had to be placed in the hospital ground to house the patients, most of whom are employees of the company'.[33] A larger hospital was completed in 1899.[34]

By 1911 the population was 3827, with a persistent imbalance in the ratio of males (2158) to females (1669). This population did not contain the diverse mix of ethnicities that characterised some other mining towns: it was overwhelmingly people from Tasmania (2854) and Victoria (490).[35] In 1921 the Queenstown municipality had a population of 3216, with less than 2 per cent born outside Australasia and the British Isles. As Townsley noted in his history of Tasmania, it was 'commonly said that Queenstown was a suburb of Melbourne', not of Hobart.[36] This related not only to the origin of migrant populations, but also to the strong links between the Mount Lyell Company and Melbourne, its head office from 1915 until 1961. This was an overwhelmingly Anglo-Celtic population, most of whom worked for the Mount Lyell Company, or in an associated industry, or were dependent on the wages of a Mount Lyell employee.

Working conditions before World War I

The Mount Lyell mines, the smelter and the associated transport and office operations were complex. After 1920, the company used modern transport techniques, including electric or hydraulic machinery, slowly reducing the amount of hard physical labour needed. The form of employment could vary too. Charlie Fox notes that from 1910 to 1920, 15 per cent of all miners at the Mount Lyell mines were on contract, while at the North Lyell mine, from 1908 to 1920, 32 per cent were on contract. The employment relationship chosen was the one thought best able to maximise production and reduce costs. For example, workers in large open-cut areas were paid a lower rate and could be readily supervised by a foreman, whereas specialist miners working underground were often without supervision – they were on contracts tied to production amounts, to increase their efforts.[37]

In October 1912 42 men were killed in the North Lyell mine disaster. Like the Mount Morgan fire of 1925, there were contending versions of the events leading up to the fire, which broke out at the 700 foot level of the North Lyell mine, and again like Mount Morgan, the disaster came after industrial trouble in the preceding 12 months. Fox reports that relations between the company and the union around the time of the disaster were 'tense and bitter'.[38] Fatalities continued in the 1920s and 1930s, usually the result of rock unexpectedly breaking away from the face of the open cut or underground falls in the stope or production area of the mine.[39] Other workers were killed by machinery, especially rail or truck accidents.[40]

The fate of the company dictated the economic fortunes and general social vitality of the town. The Mount Lyell Company was in a very strong position on the eve of the Great War. The average annual price of copper on the London Metal Exchange rose from £60 per ton in 1914 to a wartime high of £140 in 1916, and production increased accordingly. The company took a snapshot of wages in 1924 which provides an indication of the relative standing of workers in income terms. The basic wage rate for AWU members was £3/14/10 per week; carpenters secured £3/15/- per week; timber workers and engine drivers received £3/17/- and £4/0/2, respectively. As noted, Mount Lyell also offered contract work to some miners.

After the war copper prices declined markedly, and the early 1920s were difficult years. These economic conditions – declining prices, a recession, and increasing industrial conflict – were a major challenge.

Both the Mount Lyell Company and the Mount Morgan Company campaigned to reduce award wages by 20 per cent.

Unlike the Mount Morgan Company, however, Mount Lyell survived the 1920s, even ending this difficult decade in a stronger position. A new refinery was erected at Queenstown, enabling the company to further refine their concentrate to produce cathode copper. This highly refined product used to be refined by the Electrolytic Refining and Smelting company (ER&S) at Port Kembla, so once Mount Lyell had their own refinery they were able to reduce freight costs and refinery costs, though the refined cathode copper still needed the electrolytic process at Port Kembla. The loss of the Mount Lyell concentrate work produced consternation at Port Kembla, severely affecting ER&S on the eve of the depression. On many fronts the actions of the Mount Lyell Company contrasted with those of Mount Morgan. The Mount Morgan board vacillated on major investments, on drilling for more ore and on developing newer more efficient production processes, but Mount Lyell pushed ahead in all these areas.

The staff at Mount Lyell were a distinctive presence in the town. In 1920, when the workforce numbered 1481, there were 231 staff employed across the four main areas of the Mount Lyell operation: the reduction works, the mine department, the railway department and the general staff. There were 76 technical staff, 72 clerical staff, and 83 foremen and gangers. Being on staff meant a certain social status and regular income, as most staff were employed on monthly contracts (yearly contracts for the senior staff). The staff were often long-serving and fiercely loyal servants of the company. In 1929, 70 staff members had an average length of service of 15 years. The staff enjoyed access to a staff provident fund after 1924, to which both the staff member and the company contributed 2.5 per cent of their salary, with older employees securing an initial top-up. The company estimated that the staff provident fund cost them £3000 to £5000 annually, with an initial start-up cost of £22,000.[41]

Despite these benefits, the staff were not immune to major downturns. By June 1922, after one of the most difficult 12 months for the company since its formation, the daily workforce dropped to 818 and there were only 103 staff. There were retrenchments in all areas, with the largest cuts being among the foremen in the mine department, whose numbers dropped from 44 to just 17 by June 1922.[42]

The late 1920s

After the early 1920s, mine production recovered. By 1928 the company mined over 120,000 tons of ore; 20,000 tons more than the previous year. High-grade ore was sent to the converter plant, while lower-grade ore was sent to the concentrating mills. In September 1928 the new North Lyell tunnel was completed, joining the Queenstown side to the Linda Valley side. As mine historian Tony Weston noted, the tunnel led to a 17 per cent reduction in mining costs – from 22 shillings to 19 shillings per ton – in the first year alone.[43] A new refinery was also finished by mid-year. This allowed the partially refined material from the converter plant to be further refined on site rather than shipped to ER&S at Port Kembla. Despite the difficult economic times and the new investment, the company made a net profit of £200,000 in 1928; in the same year, Mount Morgan foundered and went into liquidation.

The North Lyell tunnel was not just an important breakthrough for mining efficiency; it also changed the demographic mix between Gormanston and Queenstown. Many miners lived in Gormanston, so they didn't need to make the long walk over the mountain to the North Lyell mine site; once the tunnel was completed miners could also be transported to work from Queenstown via an underground rail system. The company transferred the Mining Department to the Queenstown side.[44]

The strength of the mining business, and projects such as the new refinery and the North Lyell tunnel, provided additional male employment. By 1933, despite the depression, the population had increased to 3990, with males (2205) clearly outnumbering females (1785). There was also an increase in European-born workers, though the numbers were still small. The Census obscures the true numbers in these groups, since many came and went according to labour demand and were not captured by the Census cycles.

In terms of workforce structure the numbers were typical of a major mining town, but they also show that many 'mining' towns also had important 'industrial' functions in the area of milling, processing, smelting, and in Queenstown's case after 1927, refining. In 1933, there were 915 males engaged in mining, 297 in manufacturing, construction, repair work and utilities, and 72 in land transport, principally the railway. There were small numbers of females, with only 8 females listing occupations in the 'industrial' category. While forestry remained an important activity in the fringe settlements such as Lynchford, there were only 18 workers in this industry in Queenstown in 1933.[45]

Women were more strongly represented in the town-based occupations. The commercial and finance sector offered employment to 125 males and 46 females. The 'public administration and professional' category was also dominated by females: 43 females and 31 males. The domestic service category was another segment of the workforce where females (86) outnumbered males (30). At Mount Lyell, labour demand remained firm. But not all shared in opportunities for employment, local attachment, and the prosperity that characterised the good times.

Exclusions

Central to the pioneering ethos on the west coast of Tasmania was the idea of white explorers, prospectors and timbergetters battling and mostly overcoming the harsh environment and inhospitable terrain. The dominant historical narrative of the region is of conquering waves of white settlement.

About the Indigenous presence in this land there is a strong silence, underpinned by an assertion that Indigenous people did not occupy this region. The archaeological and ethnographic evidence reveals cultural groups and clans in the northwest and in the southwest, with

Macquarie Harbour an important marker between the two territories.

The archaeological evidence shows that Aboriginal people inhabited the southwest from at least 36,000 years ago.[46] The Queenstown region is ringed by a number of archaeological sites that show the complexities of Aboriginal use and occupation of this area, but the overwhelming focus on white land use and its mining geology has obscured this. Aboriginal people were forcibly removed or excluded from their land in the 1830s, four decades before town development began on the west coast.[47]

Even in this environment, Aborigines were still present: as figures represented in school presentations and public occasions. In 1922 the South Queenstown State School staged a pageant illustrating:

> the development of Australia from 1770 to the present day. The first act illustrated the aborigines [sic] and the landing of Captain Cook; the second act showed Australian industries; and the third act the products of the present day.[48]

Queenstown was similar to towns and cities throughout Australia in this, as similar tableaus used representations of Aboriginal people to measure the progress of white settlement, casting the Aborigines as figures that belonged to the past.[49]

Apart from the exclusion of Indigenous people, there was also a more contemporary exclusion manifest on the fringes of Queenstown society. The boundaries of this closely knit Anglo-Celtic community were occasionally policed in a vigorous, and even violent, fashion.

One example occurred in the mid-1920s. Many mining and industrial towns experienced an influx of European-born workers in this decade. There were approximately 50 Italian miners in the Queenstown area in the mid-1920s, most living in Linda Valley near the North Lyell mine, and some in Queenstown in Preston Street. Tensions between white miners and Italian miners built up through 1925 and came to a climax in 1926. In 1925, one white miner, Albert Quinn, had been killed, and another injured, in an altercation with an Italian miner.[50] After this incident, there was considerable ill-feeling and racial violence, with *The Mercury* commenting that the 'larrikins at Linda have again been molesting the Italians'.[51] Some newspaper reports suggest that many Italians moved into Queenstown after these attacks because it was not safe to stay in Linda.[52] The tension was further increased when the court in Devonport found Filippo Farajoni not guilty of the murder of Albert Quinn by reason of self-defence. The court heard that after an earlier altercation at the Linda Hotel, Farajoni was threatened by Quinn and other men who had surrounded his house. Farajoni was shot at twice with a shotgun, the second shot injuring his arm. At the trial Farajoni showed evidence of his injured arm and pellets that had been dug out of his skin.[53]

In 1926, the 'Italian camps' which fringed the North Lyell mine site near Linda were attacked. The Mount Lyell Company instructed the industrial officer to induce the police to put pressure on the Italians concerned (eight in number, all from southern Italy) to move, but the local Sergeant was reportedly unconcerned. The Italians were induced to leave by an offer from the state government to pay their fares to Melbourne. The company reported that approximately a dozen Italian men left the district, leaving 42. The general manager of Mount Lyell reported to Melbourne head office:

> The evidence at hand is that the older and more reasonable men have little objection to working with any of the Italians, although they prefer the Northern type. The trouble arises from the irresponsible larrikin type such as were involved in the Quinn affair, who let no opportunity go by to irritate and annoy the Italians, especially that particular group which also more or less associated with the Quinn affair through the occupation of huts in the vicinity of the happening. These Italians moved to Queenstown after the Linda happening, thus giving further opportunity for irritation on the long trip from the North Lyell Mine to Queenstown.

The general manager concluded that 'Italians naturally resent such treatment and retaliate in their own excitable way.'[54]

Unlike Broken Hill, there were no ethnic community organisations – such as churches and clubs – even during the 1920s, when numbers of Italians were at their highest. This absence may have encouraged violent behaviour, since the young male Italian workers were more easily excluded and demonised. Also, their residence near the isolated North Lyell mine site made them more vulnerable. The movement of Italians back to Queenstown after the violent events of 1926 indicates that racial exclusion was not tolerated or encouraged in the town itself, but could flourish in the more isolated outlying settlements.

By 1933 the number of Italians had declined. Males born in Italy dropped to 23 and females to just 2. There was not so much an Italian community at Queenstown as a floating population of young men looking for work at the mine.[55] The decline continued into and through World War II, as Italians moved on or were interned. By 1947 there were only 8 Italian-born males and 4 Italian-born females in Queenstown.[56]

Company and community relationships

For those within the acceptable boundaries of local Anglo-Celtic society, the Mount Lyell Company dominated. There were strong company town characteristics. The Medical Union, for example, was a major co-operative effort, and on the face of it appears an exemplar of the kind of cross-class localist effort that was so important in establishing infrastructure and services in new communities. The Medical Union, however was sponsored by the Mount Lyell Company. Robert Sticht was a strongly paternalist manager who opposed unions representing the workforce. According to his biographer, to 'counter what he [Sticht] considered to be the displaced loyalty of the miners, he established and became president of a medical union which offered first-rate services'. Such an interpretation is given stronger credence by the appearance of the Medical Union in 1897, one year after union activity on the mining field, led by the Victoria-based Amalgamated Miners' Association (AMA).[57] Both Sticht (general manager from 1898 to 1922) and his successor, R.M. Murray (1922–44) served as patrons of the organisation.

At Mount Lyell, welfarism and company benevolence had their origins in the industrial battle between the management and unions for the hearts and minds of the workforce, but was most clearly shaped by the company's need to attract and retain labour.[58] Queenstown was an isolated labour market, with the only access before the 1930s being the company railway from Strahan. It was a cold and unforgiving place. Charlie Fox sees the company's welfare strategy as a pragmatic response to labour supply and labour control problems. It was instituted in 1913, and after the Great War it was more formally linked to a broader 'welfare movement', often drawing on the wide-ranging experience of the Collins House groups at Port Pirie, Port Kembla and Electrolytic Zinc (at Risdon).[59] In 1939 a Mount Lyell company publication claimed that '[t]he company provides such comfort as is reasonably possible'. Working hours were 7.30am to 4.00pm. There was a change house accommodating 312 men – keeping clothes dry was a key issue. Cribs were constructed at all levels of the North Lyell and Crown Lyell mines. These were waterproof, electrically lit, and some had a heated plate for warming billies. North Lyell and Crown Lyell also had a change house for 400, with individual lockers.[60] Employees received annual leave, and had the option of taking holidays at Strahan in company-owned cottages from 1934.

The central role played by company staff at both the workplace and in broader civic and cultural organisations shaped a split social structure, with staff and their families forming a middle class, while Mount Lyell workers and their families constituted a working class. Staff families were closely connected to St Martin's church, while many working families, and labour movement leaders, were Catholics. The Australian Labor Party was as dynastic as the senior staff at Mount Lyell. Victoria-born James McDonald, for example, served in the Legislative Assembly (1915–16) and later in the Legislative Council (1916–22, 1928–47). In the mid-1920s he was AWU organiser for the west coast. His son, John McDonald, was born in Gormanston in 1904 and was the member for Bass from 1934 to 1945; another son, Thomas McDonald, born in Zeehan in 1915, was member for Wilmot and Lyons from 1959 to 1969.⁶¹

Perched high on a hill overlooking the residential areas of the town, the general manager's house, Penghana, was a symbol of company dominion over the town. Its interior also spoke of the inner life of the general manager: an extensive collection of rare and antique books graced the walls of nearly every room. Robert Sticht's most prized possessions were in his den. It was jealously guarded, and Sticht invited only select guests to view these items. His friends and colleagues included the US ambassador, senior company directors, academics and booksellers.⁶² His professional domain was the company and its workforce but his own personal networks extended well beyond Queenstown and the west coast. Under the leadership of R.M. Murray from 1922, Penghana played a more familiar role. Like Carlton House at Mount Morgan, Penghana was the site of local cultural events and fundraising efforts, with St Martin's again favoured. By the 1930s, the Murrays were hosting regular meetings of the St Martin's ladies guild at Penghana.⁶³

Despite the Mount Lyell Company's dominance, there were elements of local life that remained beyond their control. Union organisation dated from 1896, when the Victorian-based AMA established branches in Queenstown and Gormanston, with Zeehan as their base. The AMA became the Amalgamated Mining Employees' Association in 1910; the Gormanston and Queenstown branches agreed to amalgamate that year.⁶⁴ The AWU became the major general union at Mount Lyell in 1918, matching the situation at Mount Morgan and Mount Isa but in strong contrast to that in Broken Hill. This union sought to manage industrial campaigns through the state-sponsored arbitration system.⁶⁵

The unions at Queenstown were indicative of the strength of working-class organisation. At Queenstown and elsewhere on the west coast, large groups of workers clustered in the mining industry gave organised labour a presence it did not have in any other region in Tasmania. Historian W.A. Townsley argues that the origins of the Tasmanian Labor Party can be found in the west coast mines, in particular in Queenstown.[66] Visits by high-profile labour organisers such as Ben Tillet gave the movement a great boost. The events of 1912 – the disaster at the North Lyell mine and the labour militancy – also showed the vigour of the labour movement. The company's new welfarist strategies were a response to this militancy as well as to problems such as labour supply.

Unions co-operated with the company in the new welfare programs of the 1920s and 1930s, and as joint sponsors of community projects. The AWU, in particular, took on a particular character in relationship to the large single employer. At workplaces throughout Australia, including at Queenstown, the AWU became a moderate union, supporting the continued operation of the mine as much as the rights of workers. Where large numbers of workers were dependent on one employer, the AWU often identified the continuation of jobs as the overriding industrial priority. As one former employee who was active in union politics recalled:

> Then during the '80s and '90s, it seems a bit strange to say – well it's not strange, it's the way the industry was – the price dropped and all these conditions, well not all, but a few of these conditions we had, I like to think of the word we earnt, we gave – well I won't say we gave them back, but we gave them back to keep jobs and keep the mine going for another few years.[67]

This strategy was also underpinned by the crucial role that the AWU played in state Labor governments, as the ongoing viability of the west coast mines was vital to the state's economy and thus the government's electoral fortunes. By the 1950s, despite occasional forays into broader issues such as the setting of local rates and the provision of housing, the peak union body, the West Coast Trades and Labour Council, focused on negotiating industrial agreements with the Mount Lyell Company.[68]

Making a town – postwar

By 1947 there were 4017 people living in Queenstown, with males (2134) outnumbering females (1883). There were 829 males in the 'mining and quarrying' category of the Census and 18 females. The 198 men employed in the mills, the smelter and the refinery were covered in 'manufacturing'; there were very few women in this and similar categories. Once again, in the town-based occupations, numbers of males and females were more equal, with 78 males and 65 females in commerce. Females (65) outnumbered males (38) in 'public administration and professional activities', and in 'public amusements, hotels, cafes and personal services' (66 females and 36 males).[69]

The West Lyell open cut, which had opened in 1935, was the main source of mined ore, as new equipment such as mechanical shovels and 19-ton semitrailers slowly changed the nature of the workplace. The Royal Tharsis mine was the principal source of underground ore. After 1947, North Lyell was not worked, owing to labour shortages. In 1950 the company moved a record 2.24 million tons of material, and produced 8072 tons of blister copper, almost three times more than the previous year.[70] The following year saw a profit of £195,829 despite a severe labour shortage.[71]

Profitability and continuing strong demand for copper created labour demand. By the 1961 Census the population had increased to 4601, but much of this growth was in young men working at the mines. The mining workforce had increased significantly over 1947 figures, to 1300 males and 66 females.[72] Family and chain migration (where migrants from one town or region in a foreign country follow others from that place to a particular place in the new country) ensured that the newcomers of this period were not so different from the earlier Victorian and Tasmanian migrants. By 1961, 8 per cent of the local population had been born in Europe (despite major demographic changes throughout Australia, with waves of European migrants) and only a handful of residents had been born in Asia.[73]

For a mining town, the rates of home ownership were quite high, partly because the company operated a rent-purchase scheme for employees.[74] Janice Redman grew up in a three-bedroom house in South Queenstown. There was an open fire in the loungeroom, a wood stove in the kitchen and the hessian-lined walls were plastered with glued newspaper, then painted.[75] Dorothy Brown grew up in the 1940s in a three-bedroom house which her father, a painter and then smelter foreman, was paying off through the rent-purchase scheme.

In 1947 there were 927 dwellings in Queenstown, of which 905 were occupied residences. Of these dwellings, 405 were owned outright, 292 were tenanted, 157 were being purchased through a mortgage, and 26 had 'other' forms of occupancy or were not stated.[76]

Queenstown was divided into three distinct suburban areas: South, Central and North. South Queenstown had its own small commercial area, and schools. Janice Redman recalled that:

> we had everything we needed down South – we had shops, we had a school, we had a kindergarten. And it was rare, we come up the street very much [to Orr Street]. It was a big deal for us to actually... get in the car or walk up to the town because we had everything we needed down South.[77]

These kinds of suburban divisions were ritualised into sporting rivalry between the Australian Rules football teams. While Queenstown may have appeared a united locality to outsiders, internally there were finely graded differences based on geography and class. In 1899 three teams competed in the Australian Rules league: Queenstown, Smelters and Flux, the latter a team based around the flux quarry in North Queenstown.[78] By the early 1900s a team from Linda joined the local league, and newspaper reports also refer to a Railway Team.[79]

The Queenstown Football Association was formed in 1924, and by the 1930s the major teams were City, Mines United, Gormanston, Lyell, Smelters and Railway.

The importance of Australian Rules in Queenstown again reveals the influence of Victoria. Games were organised against other west coast towns. In 1926 Queenstown City won the West Coast Premiership and defeated all visiting teams, including those from Burnie, Waratah, Zeehan, Rosebery, Gormanston and East Zeehan.[80] Fierce regional rivalry helped overcome internal divisions that developed during the season.

By 1961 Queenstown had a population of 4591 (with 2518 males and 2083 females). The gender imbalance indicates the strength of the mining industry, and the jobs available for men in particular. In that year there were 1356 workers in the mining industry: 1300 males and 56 females. The population remained overwhelmingly Anglo-Celtic. The Anglican community remained strong, with a 37 per cent professing adherence to the Anglican faith at the 1961 Census; 32 per cent were Catholic; 10 per cent were Methodists; 5 per cent were Presbyterian; and 1 per cent were Salvation Army.

Company and community relationships – postwar

In the decades after 1945, the Mount Lyell welfare policies matured into a wide-ranging program of workplace and community support. They were as much a part of local culture as they were labour management strategies. The main supermarket was the Mount Lyell Supply Store, and there was a company butcher in Gormanston. The company had built many of the houses in town, including the new estates on the northern edges. Mount Lyell encouraged Gormanston-based employees to move from their old company homes to new company houses in North Queenstown, aiding and abetting the decline of Gormanston and the growth of North Queenstown.[81] As Dorothy Brown recalls, 'that was when Gormanston just sort of went down'.[82]

The economic and commercial dominance of Mount Lyell still played out socially, with the senior staff socialising at the Gentlemen's Club, the Empire Hotel, and the Golf Club, and the waged workers keeping to the town's other hotels. As one resident noted:

> Everybody knew their place and accepted it. Because the Mount Lyell [Company] just did so much. The general manager was like the King of Mount Lyell.[83]

This was made clear when Penghana was opened to the public (for weddings, public occasions and visits) – former Mount Lyell workers were very reluctant to enter the building. Many stayed on the lawn while their wives walked through.

Senior staff, the elite in the town, were still likely to be members of St Martin's Anglican church, and were often active in educational, cultural or council matters. The senior leadership of the company operated in an almost dynastic way (with a few exceptions), and senior positions often passed from father to son. A few examples will illustrate the general point. W.L. Hudspeth was a long-serving company secretary. At the wedding of Betty, Hudspeth's eldest daughter, in 1938, were the cousin of the bride, Hugh Murray, and her brother Geoffrey Hudspeth, who 'officiated as best man and groomsman respectively'. Both became general managers of Mount Lyell, Murray in 1949. The wedding was, unsurprisingly, held at St Martin's, and the reception was at the staff hotel of choice, the Empire Hotel.[84]

Hugh Murray was the eldest son of long-serving general manager R.M. Murray. L.K. Hudspeth's son, Geoffrey, progressed through the ranks of the staff, serving as mine manager from 1949 to 1965, and returning as general manager from 1967 to 1973. Both father and son were involved in the Workers' Education Association at Queenstown. Hudspeth senior was involved in the management committee of the town library, the Robert Sticht Memorial Library, as was his wife. On the departure of L.K. Hudspeth and his wife in 1941, Mrs Hudspeth was farewelled at a gathering at the Empire Hotel with 80 ladies present. The long list of Mrs Hudspeth's voluntary activities epitomised the extensive work that many senior staff wives undertook. The list included the Red Cross, the Child Welfare Association, the Dorcas Society, the Horticultural Society, the Ladies Hospital Auxiliary, the Girl Guides, and the Robert Sticht Memorial Library. On retirement, the Hudspeths reportedly took up 'residence on the mainland'.[85]

Their son Geoffrey was not only the mine manager and a high-profile participant in Queenstown's civic organisations. In 1949 he became a councillor on the Queenstown Council. It was no easy task to distinguish between the roles of mine manager and councillor. At a council meeting in January 1951, for example, Hudspeth spoke of the difficulties the company experienced in moving heavy equipment on the Lyell Highway from Hobart to Queenstown:

> Recently we brought nine heavy duty ore-carrying trucks to Lyell. Each had to be dismantled before and re-assembled after crossing the bridge. Bringing heavy machinery across the bridge at night is a dangerous job. Workers have to rely on jacks and slings and it would be 'dead easy' for someone to be killed.[86]

The 'we' in this speech referred to the company, not the council, despite this report coming from a council meeting. Hudspeth then suggested that it was '"high time" the Public Works Department or the Government did something about the King River bridge'. That may well have been the best solution to this problem, but the blurring of the company and the council's voices is clear. Hudspeth was not the only senior staff member involved in council politics. H.G. Faull was superintendent at the Mount Lyell locomotive workshop. He was a Queenstown councillor for 34 years, including six years as warden, and a president of the West Coast branch of the Liberal Party.[87]

Hospital opening procession, 1889. Local friendly societies and trade unions, including the AMA, marched down Argent Street in 1889 to celebrate the opening of the new hospital. AMA President J. Neil was the 'grand marshal' for the occasion. Ion Idriess Glass Plate Collection, 1885-ca. 1954, NLA.

Head frames and smokestacks along the line of lode, Broken Hill, c. 1892–93. Fred Hardie, Fred Hardie G.W. Wilson collection of lantern slides, NLA.

Suburban growth at Broken Hill, 1959. By this date, municipal and resident tree planting had softened the once hard streetscape. Wolfgang Sievers, Wolfgang Sievers photographic archive, NLA.

The line of lode from the Block 10 mineral lease area, 2006. The mining and industrial workplace has been replaced by an access road and a memorial (to the left at the top) which commemorates those who lost their lives in the Broken Hill mines. Photograph by the author.

Above left View of the Mount Morgan mine and general office taken by Frank Hurley, c. late 1940s. Hurley negative collection, NLA.

Above right View of Mount Morgan mine site taken by the author in 2008. The photograph can be compared with the Frank Hurley shot from the late 1940s (above left), taken from a similar location.

'View of Morgan Street, Mount Morgan, Queensland, c. 1910.' Commercial and retail development, along with trade and traffic, are features of this photograph from the heyday of the town's prosperity. Part of Scenes from Mount Morgan Queensland, 1900–1937, England, Lee & Humphreys, 1900–1937, NLA.

The Mount Lyell smelters, next to the site of the former township of Penghana, whose faint outlines can be seen. Stephen Spurling III, Spurling Collection of Photographs of Tasmania, 1903–1930, NLA.

Streetscape in South Queenstown showing old and newer homes. Behind, low cloud hugs the top of Mount Owen. Photograph by the author, 2006.

Port Pirie Uranium Treatment Plant, 1956.
Australian Information Service collection, NLA.

This staged photograph, probably from the late 1870s, shows the proximity of the main street, Ellen Street, to the railway terminus and the harbour front. J.C Dunn's was a prominent building on the waterfront area. Samuel White, 'Album Views of Adelaide, suburbs and industrial area and countryside', NLA.

Above left Looking from the fishermen's wharf area towards the Nyrstar smelter and the prominent stack. Photograph by the author, 2011.

Above right A view of the MIM operations across the Leichhardt River from a hotel balcony in the late 1940s. The number of cars coming back across the bridge towards the township, and the angle of the sun, suggest the end of day shift. Frank Hurley, Hurley Negative Collection, NLA.

Overlooking the Mount Isa commercial centre with the Xstrata operation, including the large lead smelter stack and the stripped copper smelter stack, in the background. Photograph by the author, 2006.

The distinctive curved design of Kambalda East with the WMC operations and Lake Lefroy in the background, 1973. Wolfgang Sievers, Wolfgang Sievers Photographic Archive, NLA.

An artist's rendering of the planned WMC houses in West Kambalda, 1969. WMC Collection, University of Melbourne Archives.

'Miners finishing a shift at Western Mining Corporation's Silver Lake shaft at Kambalda in southeast Western Australia, 28 March 1969'. NLA. It has not been possible to trace the names of these men.

'Photo of Anthony Edwin Bowes Kelly and wife Mary taken at "Moorakyne"'. After making his fortune through BHP and then Mount Lyell shares, Kelly and his wife moved to 'Moorakyne' in 1901. Image courtesy of Stonnington Library, Melbourne.

1968 and beyond

Major changes to the nature of the operation came in the 1960s, after many years of growth and expansion. The Queenstown to Strahan railway was closed in 1963, as truck haulage was found to be more efficient. The refinery was shut down in 1965 and concentrate was again sent to ER&S at Port Kembla, as it had been before the refinery was built in 1929. Smelting ceased altogether at Queenstown in 1969. The mine that had precipitated downstream industrial development at Queenstown reverted to mining-only. Like Mount Morgan, the Mount Lyell mine also suffered from the decline in copper prices: they went from a high of US$2053 per tonne in 1974 to $1232 in 1975, then back to $1978 in 1979.[88]

These were difficult years at Queenstown. From 4622 in 1976, the population declined to 3721 in 1981; this was a major drop for such a small community. In 1976, 344 workers lost their jobs as major components of the mining operation, such as the West Lyell workshop, closed.[89] Underground mining returned as the West Lyell open cut was worked out, but this required more capital expenditure. The company became part of Renison Goldfields Consolidated in 1981 and struggled through the 1980s.

After the retrenchments of 1976 it was difficult to find work in Queenstown. Janice Redman returned after three years in the army and was unable to find regular paid work. She looked after her mother and father, worked in their vegetable garden, did casual cleaning in the local motels, and started researching local history.[90] Her experience was typical for local women in a town which was still dominated by mining industry employment and where 90 per cent of the employees were men.

External influences

The Tasmanian government was a key player in development on the west coast. While the Emu Bay Railway and the two mining company railways were private ventures, the government constructed the Strahan to Zeehan railway, and smaller tram lines linking the outlying mining settlements. After the Mount Lyell company opened the Lake Margaret hydro-electric power scheme in 1914, the state government also moved into power generation, damming the state's rivers and using the flowing water to spin turbines to produce electricity. By the 1920s, government provision of cheap power, through the Tasmanian Hydro-Electric Commission, was a central plank in the political economy of Tasmania, building dams and associated infrastructure throughout the state and attracting industries such as pulp and paper mills and base metal refineries.

By the 1980s some 19 industries had been established in Tasmania, and 'the Hydro' employed 4000 workers. Mainstream politicians were united in their support of the Hydro.[91] Premier Eric Reece (who held office from 1958 to 1969, and again from 1972 to 1975), was a former Queenstown miner and AWU organiser, and was widely and affectionately known as 'Electric Eric'.[92]

During the latter stages of the mine's life state support was even more evident. In 1975 the mine was given an exemption from the Tasmanian government's environmental legislation on the basis that the continued operation of the industry was more important to the state than any possible or evident environmental harm.[93] From 1977, as the company struggled, the state offered large loans on generous terms. *The Mount Lyell Mining and Railway Company Limited (Continuation of Operations) Acts* of 1985, 1987 and 1992 provided loans to the company to offset the effects of low commodity prices. In 1995 the old company and the state government signed a Memorandum of Understanding under which the company would repay $7.946 million but not the $4.3 million outstanding.[94] The Commonwealth joined in, matching state support through the *Tasmania Grant (The Mount Lyell Mining and Railway Company Limited) Act 1977* (Cth) and subsequent amendments.

Government support was indicative of Queenstown's special place in the political economy of Tasmania. The continued operation of the mine was an article of faith for Tasmanian politicians. In the face of poor base metal prices and an environmental battle on the west coast — the campaign to stop the Hydroelectric Commission damming the Gordon River — the mine's continuation became almost talismanic for those who supported economic development in Tasmania generally. In Queenstown, and throughout the west coast, the campaign to stop the dam was matched by a local (and state-wide) effort to organise working people in support of west coast development. A lobby group, The Organisation for Tasmanian Development, secured support from Premier Robin Gray, who had just won an election on the platform of 'state development', and from trade unionists such as FEDFA (Federated Engine Drivers and Firemen's Association) leader Kelvin McCoy. In December 1982, a rally in Queenstown in support of the dam was attended by 3000. Queenstown was a household word among mining investors in the 1890s, and in the 1980s it came to national prominence again, this time as a site of popular resistance to green politics. It was only from the 1990s that it became clearer that the Hydro's plans (which were ultimately thwarted by Commonwealth government intervention in 1983), as well as the Mount Lyell mine and its environmental effects, could well have compromised plans for alternative economic development, including ecotourism and aquaculture.

The end of the old company

Despite extensive state and Commonwealth assistance, the Mount Lyell Company went into liquidation in 1994. Many of the mainstays of the company's mining production had closed. The Lyell Comstock closed in 1959, the West Lyell Open Cut in 1978, and North Lyell in 1972. Cape Horn mine closed in 1987, and the Royal Tharsis mine persisted until 1991. Copper Mines of Tasmania (CMT) Pty Ltd took up the Mount Lyell lease in 1994, with mining in the Prince Lyell area commencing in 1995. In 1999 CMT was acquired by an Indian company, Sterlite Industries (India) Limited, itself a wholly owned subsidiary of another Indian company, Vedanta Ltd. In 2009–10 some 1.8 million tonnes of copper ore were mined, producing 23,160 tonnes of concentrate and employing 285 people.[95] One 2001 estimate indicated that the mine had produced 1.2 million tonnes of copper and 4 tonnes of gold since 1893.[96]

By the 1990s, it was no longer possible to imagine a viable future for Queenstown based on mining alone. Also, a wider appreciation of history and heritage conservation had been growing throughout Australia since the 1960s. At Queenstown it emerged much later, in efforts to preserve the past and use it for heritage tourism purposes. A mix of volunteer organisations and state and council-funded projects have sought to capitalise on the town's mining and community history. The most high-profile heritage development was the re-opening of the Queenstown to Strahan railway as a tourist train. Earlier efforts included the Galley Museum, established in 1985 by a former miner and amateur photographer, Eric Thomas, and still run by volunteers. The Mount Lyell Company's general office was acquired by the National Trust in 2010, and plans are afoot for further conservation and interpretation of this site.

The sulphur dioxide fumes from the Mount Lyell smelters scarred the valleys and surrounding hills. Mount Lyell and Mount Owen were slowly denuded of vegetation, the result of widespread clearing, the sulphur fumes, and bushfires. Scoured by heavy rain, the hills took on a remarkable appearance that became a symbol of local difference and identity.[97] For non-locals, these bare hills were one of the defining images of the late 20th century backlash against unfettered progress, and a powerful statement about the deleterious effects of mining on the environment. In the last ten years the hills have recovered some of their vegetation naturally. The sulphur dioxide has leached away. Other parts of the regrowth have been given human assistance.

This regeneration has had a mixed reception among locals. Many welcome the return of the trees and the greenery, but at the same time those bare hills were a distinctive part of their place. These locals grew up with them; they were a defining element of the area. One glance outside could confirm that you were in Queenstown. As outsiders marvelled or scorned the sight, locals held on to it defiantly as a marker of their identity. It was the working-class battlers of the town against middle-class outsiders and 'greenies' who had little sense of the town's heritage. This split over the image of the hills reached its zenith in the 1980s, when Queenstown was widely known as the least green-friendly town in Australia.

The end of large-scale mining has changed the social and economic fortunes of the town and its surrounding area, and has left an undeniable environmental legacy. Environmental scientists have estimated that the Queenstown smelters emitted 200,000 tonnes of sulphur dioxide during the period 1896 to 1903.[98] Later smelting processes were less polluting but emissions continued until the smelter closed in 1969. The mine deposited 97 million tonnes of tailings and slag directly into the Queen River, and 1.4 million tonnes into the King River.[99] The mining leases remain a major source of acid mine drainage into the river system, which eventually flows into Macquarie Harbour. The waters of these rivers, as one study noted, 'are typified by low pH, and high concentrations of sulphate, copper and aluminium, iron, manganese and other trace metals'.[100] A major monitoring and remediation program was established by the Commonwealth and state governments in 1994.[101] The subsequent owners of the mine, CMT, constructed a tailings dam on the Princess River to manage runoff from their new mining operations.[102]

Queenstown struggles with the social and environmental legacies of over a century of mining. The town's history is closely linked with the fortunes and policies of the Mount Lyell Mining and Railway Company. The company helped shape a local social structure which was divided between staff and waged employees. For many decades this overwhelmingly Anglo-Celtic population passed leadership positions from father to son. There was limited space for women in this place, dominated as it was by employment in mining and its associated industries. Even the local council laboured under the weight of the company's dominance. Local unions and the Labor Party established a presence, but were also influenced by familial and dynastic traditions. Working-class politics mirrored the staff politics, as influential Labor

Mining Towns

families passed leadership from one generation to the next. Luckily, the historic fabric of the town is largely intact, with the main street and surrounding areas continuing to evoke the town's past in a most powerful way.

5
PORT PIRIE
'Essentially hard and practical'

Port Pirie's grid-pattern streets map uniformity onto an unruly environment. This is dry country, starting with flat scrub, wetlands and mangroves near the coast, then stretching to undulating rich farmland in the hinterland. The southern edge of the Flinders Ranges frames Port Pirie to the north. The mean annual rainfall is 344.3mm, and it mostly falls from May to October. The mean maximum temperature is 31.9°C and the mean minimum is just under 8°C. The summers have an average of 18 days with temperatures over 30°C, and 10 days over 35°C.[1] Port Pirie sits on the upper reaches of the Port Pirie River, with parts of its commercial centre on land reclaimed from the spring tides which once regularly flooded the area. Located near the top of Spencer Gulf, 225km north of Adelaide, Port Pirie sits near other industrial port towns: the steel town of Whyalla is across the gulf, and Port Augusta, an important service centre for outback regions, is a further 100km north. The three towns are often considered together as the 'iron triangle towns'.

Port Pirie has a strong sense of identity, forged initially through competition with other port towns, and later by a defensive coalition with them: South Australia is a highly centralised state, and resources and development have focused on Adelaide or on rapidly growing sites such as Whyalla. The town grew out of an existing hinterland, which meant that a well-connected and politically effective local middle class was able to lead successful campaigns for transport infrastructure and industrial investment in the 1870s and 1880s. Later,

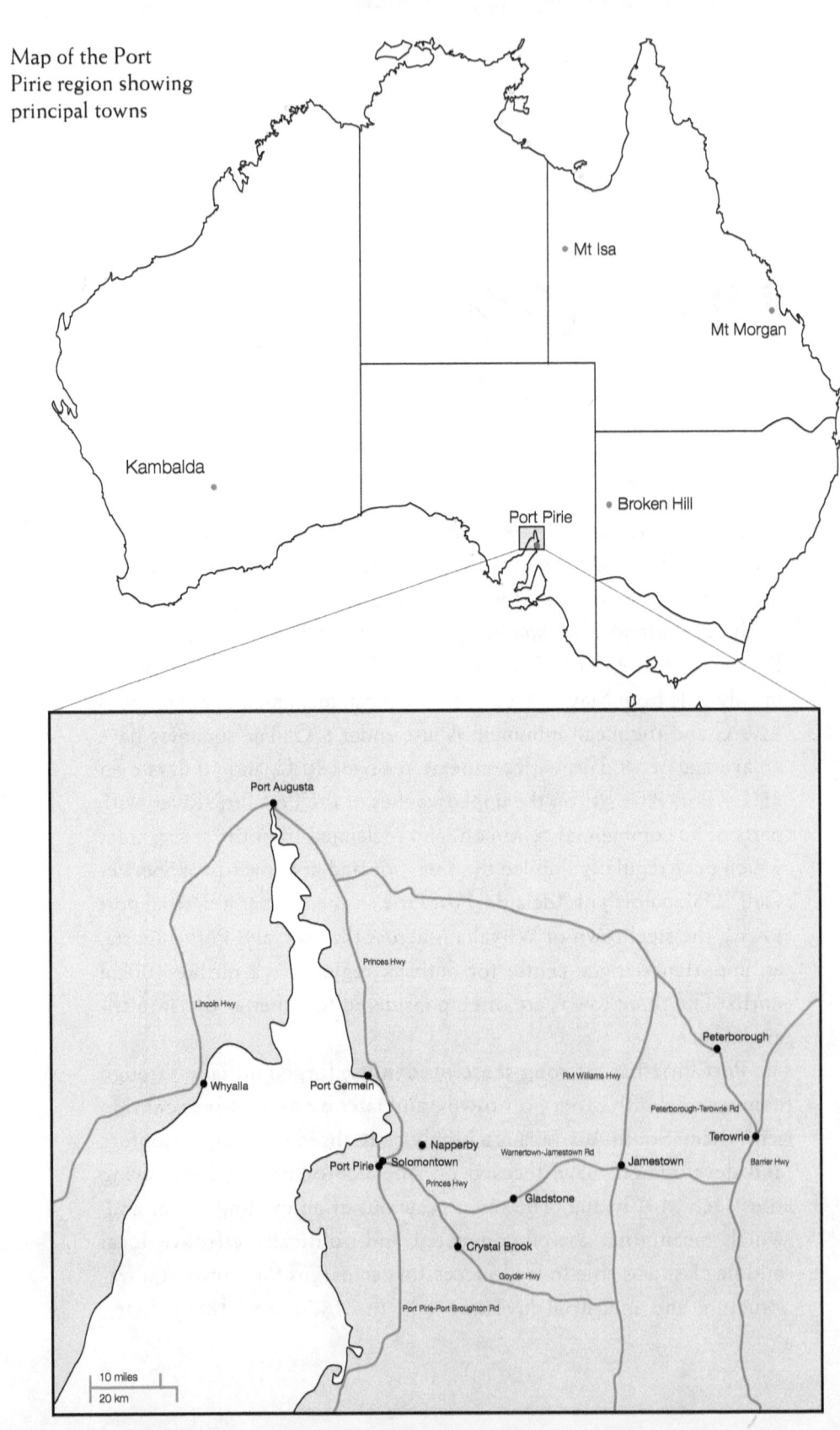
Map of the Port Pirie region showing principal towns

ethnic diversification and conflict, and occasional class mobilisation, complicated and undermined the sense of unity. One large employer, Broken Hill Associated Smelters (BHAS), sought to establish control over town society from 1915, and achieved considerable success on the industrial and cultural fronts. The diversity and complexity of the labour market, however, meant that the level of BHAS control never matched that exercised at Mount Isa or Kambalda.

Beginnings of an industry

Port Pirie began as a shipping site in the 1840s, when the mid-north of South Australia was opened to pastoralists. The place got its name in 1846, from the schooner *John Pirie*, which had been working the Spencer Gulf area transporting sheep for pastoralists. The master of the *John Pirie*, Captain Thomson, navigated the creek entrance, with its sandy shore, mangroves and tidal flats, and his efforts were reported by the Adelaide papers in March 1846.[2] A private township was planned in 1848 (for what later became Solomontown), but few blocks were purchased and there was no development at the site. By the late 1860s there were only three woolsheds on the sandy shore.[3]

Given significant encouragement by the Selection Acts in the late 1860s, the pastoralists were soon joined by wheat farmers. South Australia became a major wheat producer in the 1870s through state support and a series of good seasons: wheat joined and then surpassed wool as the port's main export. The area of wheat under cultivation in South Australia grew from approximately 605,000 acres in 1870–71 to over 1.7 million acres in 1880–81.[4] This propelled town growth across the region. Port Pirie was surveyed and subdivided in 1871, this time by the government, and a municipality with elected councillors handling local matters such as roads and public health was formed in 1876. By 1881, the population was 901 (514 males and 387 females) living in 251 dwellings.[5]

Local Indigenous people, the Nukunu, were profoundly affected by these sweeping changes to regional land use and the consequent depletion of their natural resources. There was violence on some pastoral stations. In 1854, at Crystal Brook station, local Aborigines were killed by Peter Ferguson and his stockmen for taking cattle.[6] A reserve was established at Baroota, 35km north of Port Pirie, sometime before 1911. It included a church, houses and a farm. Norman Tindale collected evidence of Nukunu language and traditions at Baroota, as did Port Germein schoolteacher Roy Parkes. This material was later used by Louise Hercus to record the language.[7] She gives the Nukunu word for Port Pirie as Thalpiri. The written records are mostly silent on the fate and experiences of the Nukunu. Port Augusta today has a large Aboriginal population, so consolidation may have occurred towards that town. The Nukunu remain a significant absence in this chapter, and future research will need to remedy this.

The 1880s marked an important shift in the economic focus of the port's activities as ore from Broken Hill began arriving by rail at the town's wharves. Some Broken Hill-based smelting companies shifted their processing plants to seaboard locations in the 1880s and 1890s. The British Blocks Company began operating furnaces at Port Pirie in 1889. This led to a major increase in the population and in economic activity. By 1891 there were 4006 residents (2342 males and 1664 females) and 725 houses.[8]

Broken Hill Proprietary Limited (BHP) initially leased some furnaces from the British company, and from 1892 began setting up their own large works on site. By 1896 there were two smelting companies, and various coke-making facilities in Port Pirie, employing around 1000 men in total.[9] BHP, convinced of the advantages of smelting at a seaboard location, then decided to shift all its smelting operations to Port Pirie. By 1904, BHP employed an average of 1070 workers for the year, and by 1913 this figure had increased to 1500; other related engineering employment appeared too, including Forgan's foundry, established in 1893.[10]

The region

Adelaide dominates South Australia. In 1930, it had a population of 310,526, which was 54 per cent of the state's population. This proportion steadily increased throughout the 20th century, with a major increase between 1933 and 1947.[11] The urban centres that developed outside the metropolitan area were either mining, rural service, or coastal service towns. There was an important phase of town development with the copper discoveries in the 1840s, and again in the 1860s. Burra, in the Clare Valley, developed from the 1840s, with a large migrant population from Ireland, Wales and Cornwall; the 'copper triangle' towns of Wallaroo, Moonta and Kadina on the Eyre Peninsula emerged in the 1860s. By 1961 there were only four non-metropolitan towns in the state with a population close to or above 10,000: Port Pirie, Port Augusta, Mount Gambier and Whyalla.[12]

Port Pirie's strong rural hinterland is in contrast to Queenstown, Broken Hill, Mount Morgan and Mount Isa, which all owed their existence solely to the discovery of a rich orebody. First wool, then wheat, was grown in the rural areas around Port Pirie. After two years of agitation and advocacy, the town secured a narrow-gauge rail line to Gladstone, 40km away, through the *Port Pirie Railway Act* of 1873. This was a crucial victory, beginning a long line of successful bids for state support and private investment, which culminated in the narrow-gauge line to Cockburn, near the SA/NSW border, in 1887 – this took advantage of the Silverton and Broken Hill trade, and the new Port Pirie smelters in 1889.

Port Pirie's hinterland expanded through state policies on land settlement (particularly the collection of Acts generally known as the *Strangways Act*, dating from 1869), agricultural development, and investment in transport infrastructure. By 1876 Jamestown had a population of 1216, Crystal Brook 954, and Napperby 540. Petersburgh (later Peterborough), some 100km east of Port Pirie, also grew steadily. Petersburgh was linked to Jamestown by rail in 1881 and in turn was linked to Adelaide through a line from Terowie. It became a key area for railway employment.[13]

The region was full of town-based competition, and Port Pirie was embroiled in often bitter competition with rival port towns such as Port Germein, and with Port Adelaide itself. Port Germein, a coastal settlement 25km to the north, had none of the disadvantages of the river port, and had prominent supporters in the 1870s, including the president of the Marine Board.[14] But the extensive shallow water approaches meant that a lengthy jetty was the only way for vessels of any reasonable size to berth. A quick trip through present-day Port Germein, with its population of 250, reveals who won the final battle for the hinterland and Barrier trade. From the now disused jetty at Port Germein, the tall stacks of Port Pirie are visible to the south, and at night the lights of Pirie sparkle on the southern horizon.

Port Adelaide was a strangely ineffective rival for the Barrier trade. Some Adelaide merchants resisted development at Port Pirie, but others set up branch offices there. Still others were Broken Hill investors, who stood to gain from Pirie's growth no matter what its effects on Port Adelaide were. Ore from the Barrier mines was carted to Terowie, 100km east of Port Pirie, then railed to Port Adelaide. Two smelters were constructed in the Port Adelaide area, close to the northern railway, receiving ore from the Barrier, and from Queensland and other parts of New South Wales. However, as freight to Port Adelaide cost more than to Port Pirie, these smelters were not competitive with any Pirie-based operation.[15]

Port Pirie exemplifies the connectedness between the mining sector and the emerging industrial manufacturing sector. The profits from Broken Hill underwrote industrial expansion at Port Pirie just as Broken Hill ore fed its furnaces. If there was supply trouble or industrial action at Broken Hill, Port Pirie was directly affected. During the Broken Hill lockout from January to May 1909, Len Zubrinich was working for a blacksmith in Port Pirie. Interviewed in 1982, he recalled how the lockout 'affected everyone and it cut down a lot of enterprise in South Australia'.[16] In August 1919 production at Port Pirie again halted as the effects of the Great Strike at Broken Hill were felt. So it is no surprise that the representatives of both capital and labour sought to influence the town's politics and its labour movement. For the labour movement at Broken Hill, Port Pirie was strategically significant, and a source of support and comradeship during disputes. Likewise, for the directors and managers of the Broken Hill mines, especially after 1915, Port Pirie became a place where experiments in new labour management techniques could be trialled without the criticism and resistance of the Broken Hill unions.

The connections with Broken Hill have been outlined in Chapter 1 and above, but these connections went well beyond trade, economic and financial links, and persisted throughout the 20th century. The relationship between the Barrier and the Port Pirie labour movement was a complex one, with instances of great solidarity and fraternity and instances of conflict and bitter division. This has been compounded by the fact that the workers of Broken Hill are more strongly represented in the written record, so Port Pirie labour movement tales have often been written by people from other places. Yet there were other dimensions to the relationship: a shared passion for Australian Rules, and Port Pirie and the region being a holiday destination for Broken Hill workers. Many Broken Hill respondents had knowledge of Port Pirie, though seaside holidays or family relationships. Port Pirie was a crucial gateway town to the Barrier from the beginning of the silver rushes in the 1870s. After 1888, the rail line solidified this relationship, becoming a powerful conduit that brought more than just Oregon timber to Broken Hill.[17]

Finally, as befitting a connected port town which performed an important transport function, from 1937 the port became a crossroads for three railway systems using different gauges.[18] The narrow-gauge lines were the cheaper freight lines running to the hinterland and beyond; the standard-gauge line was the transcontinental line that traversed the length of the country from 1937; and, also from 1937, Port Pirie was connected to the newer broad-gauge lines that originated in Adelaide. Port Pirie was a 'break-of-gauge' station with complex yards and maintenance facilities, so the South Australian Railways provided a significant amount of employment locally and in the region. The rail lines still punctuate the urban fabric of the town, crossing thoroughfares such as Ellen Street near the grain terminal.

Making a town

The commercial centre was built, just beyond the high-tide mark, from the early 1870s. The first commercial premises, Horner & Co., were registered in 1872; the Port Pirie Hotel was constructed in 1874. Numerous buildings were set on piers to escape damage from spring tides or tidal surges. In a good season, the commercial centre was an impressive sight. Shipping agents, importers, merchants and wool buyers were joined by all the trades necessary for transport and shipping, such as carters, bullock teams, farriers, wheat lumpers and trimmers. A flour mill opened in 1874. From December, when the wheat crop was bagged and sent to Port Pirie, the main street, Ellen Street, was jammed with bullock teams. On the western side of Ellen Street, the river waterfront bustled with crews and cargoes awaiting loading. In 1876 the railway line from Jamestown was completed, running directly down the centre of the street. Slowly rail carriages replaced the bullock teams. By the 1890s, the town had an established commercial and industrial centre and a basic street grid laid out behind the reclaimed waterfront, which was lined with parallel wooden wharves.[19] The mangrove and low-lying commercial and suburban areas that were reclaimed often used slag from the smelter. The parks too were lined with slag. This would later turn out to be a major environmental hazard, since the slag contained lead and other heavy metals.

Port Pirie had developed rapidly, without much basic infrastructure. One major problem was the lack of an adequate water supply. Parks and gardens were difficult to maintain, and the port's hot summers took their toll. Another problem, perhaps even more significant, was noxious fumes, predominantly sulphur dioxide, a by-product of the smelting process. In 1923 the town gardener, a BHAS employee, complained that:

> The smoke fumes from BHAS Prop. Coy. Ltd's Works have caused considerable damage to vegetation of all kinds during the past year. It is very difficult to fight against this nuisance.'[20]

When asked what Port Pirie was like, Len Zubrinich replied that sulphur used to fall 'all over the place, all over Pirie'.[21]

It is little wonder that in 1899 one visiting journalist noted that 'there is no greenery or scenery in the town, no parks, gardens or verdure of any kind. Pirie is essentially hard and practical.'[22] It was seen as a rough, 'uncivilised' place that lacked many of the necessary features of modern life. In 1923 George R. Brickhill, the editor of *The Recorder*, one of the two local newspapers, lamented that:

> the town is like a business without capital. It has to struggle along from hand to mouth, unable to afford improvement and with little hope of establishing beauty and comfort.

For Brickhill, improvements were essential if Port Pirie was to change from 'a mere money making place into a real home town'.[23] The town's rough-and-tumble male social life – there were, for example, 340 convictions for public drunkenness in 1909 – and its basic infrastructure were a cause for concern for the company.[24]

Attempts by BHAS to encourage a better 'class' of workman were targeted at making Port Pirie more amenable to married men. By 1921 the population had grown to 9801. Port Pirie was one of the few industrial towns where females outnumbered males: in 1921 there were 4953 females and 4848 males, despite there being large numbers of single men, or men travelling away from their families, living there. This relates to Port Pirie's origins as a port town with strong links to its rural hinterland. There was no sudden increase in population as occurred on the mining fields; there was steady growth and some migration from the hinterlands. This made for a more balanced gender ratio.[25]

Housing had distinctive local features but was also reminiscent of the miners' cottages of Broken Hill and Kalgoorlie. Early homes were made of pine which grew to the east of the township, but this resource was soon gone. The 1921 Census records 2308 occupied dwellings and 62 boarding houses, a number which must be regarded as a minimum, since the Census may not have recorded smaller, 'unofficial' operations. Those not in boarding houses, hotels, or tents in 'canvass town' in Pirie West, lived in simple corrugated iron or wood homes, often in trying conditions. There were stone houses too, usually for larger commercial or civic buildings, and for the better-off. The stone was dolomite quarried at Napperby and to the east of Port Germein. In 1921 there were 396 stone and 179 brick houses, but the majority were made of iron (864), or wood (686).

From the early 1890s, the local economy had two major centres of union organisation, the wharves and the smelters, but the smelters were increasingly dominant. The Australian Workers' Union (AWU) became the largest and most important union in the town in 1917, when it effectively took over the other major BHAS-based union, the Amalgamated Smelters' Association (ASA). The ASA was closely linked to management, but a campaign led by the Labour Council effectively isolated the branch and it was declared a 'bogus union'. The AWU also covered workers on the tugboats, the horse drivers, former members of the United Labourers Union, and Port Pirie Municipal Employees.[26] By 1917, the union had around 1200 members, with the vast majority employed at the smelters.[27] The AWU was a moderate union, committed to arbitration and loath to use industrial action.

The wharf labourers were covered by the Amalgamated Workers' Association (AWA), formed in 1885. The AWA affiliated with the national Waterside Workers' Federation (WWF) in 1902, and by 1920 the WWF had 507 members.[28] Since BHAS did not employ waterside labour at Port Pirie, this group of workers was largely untouched by BHAS labour management practices. More than any other union at Port Pirie, the WWF represented an arena of independent working-class industrial and cultural activity.

Craft unions such as the Amalgamated Society of Engineers' (ASE) and the Federated Society of Iron Shipbuilders and Boilermakers were also represented at Port Pirie. The ASE had 25–30 members in the early 1920s, and the Boilermakers had 29 members in 1920.[29]

The formation of the Port Pirie Trades and Labour Council in 1891 was followed by increasing political organisation. There was success for local Labor candidates in colonial and later state parliaments, and even in the early 1890s Labor candidates were contesting municipal elections.[30] By 1900 the labour movement was comparatively well organised; it had to be to survive an authoritarian employer in the form of BHP, but it also received support from Broken Hill workers. However, relations between these two labour movements deteriorated: first, because of a controversial return to work by the Pirie men during the 1909 lockout, and second, because of a bitter split between the Barrier Amalgamated Miners' Association and its Port Pirie branch in 1915–16.[31] But judging Port Pirie workers by Broken Hill standards would be unfair. Having seen off BHP, unionists at BHAS faced a very different employer, one with more liberal attitudes. They lacked the radical group of underground workers who gave Broken Hill a militant edge, and they were generally supportive of state-sponsored arbitration, rather than the workplace-based agreements that were formulated on the Hill. At the 1915 election Labor won Port Pirie, one of 26 seats with which it formed government under Premier Crawford Vaughan.[32]

Union and Labor Party organisation was matched in its scope and effectiveness by local businessmen. The commercial and trading businesses at Port Pirie provided a core of influential and well-connected figures who were important spokesmen for the town. This group continued to exercise some influence, usually through the activities of the Port Pirie Chamber of Commerce, despite the growth of BHAS as the dominant employer.[33] Men such as Nicolas Simons in the 1870s and 1880s, Fred Grey in the 1890s and 1900s, and Charles Geddes in the 1910s and 1920s were active in promoting the town, and in its campaigns for better infrastructure. Having a common commercial rival, whether it was Port Germein or Port Adelaide, united local business: the Port Pirie Wharf Company was an example of local businessman combining to build locally owned wharf infrastructure to rival that installed by Port Adelaide merchants.[34]

Company plans and resident experiences

The arrival of BHAS in 1915 had profound effects on Port Pirie society. Smelting had already broadened the port's labour market but BHAS had more ambitious goals. Essentially, BHAS attempted to graft a company town sensibility onto an existing port and rural service town. It was a major project, embodying elements of US and British industrial welfarism. The Port Pirie experience is also significant in that the labour management strategies developed there filtered through to other mining and industrial towns in Australia in the 1920s and 1930s. Managers at Mount Lyell, Port Kembla and Mount Morgan all showed an interest in the Port Pirie developments.[35]

The new BHAS management began by investigating the state of the smelter. Their initial impressions were unfavourable. BHP had allowed the works to deteriorate. Key Director and Collins House figure W.S. Robinson called it 'a lot of old junk'.[36] For example, ore was still transported by wheelbarrow (known locally as 'banana carts') or horse and cart.[37] On the ground at Port Pirie, the details of the new welfarist schemes were left to Gerald Mussen and Gilbert Rigg. Mussen was BHAS's industrial representative, and was later involved in industrial relations at Broken Hill, where his attempts to introduce Co-operative Stores became widely known as 'Mussenism'.[38] Rigg, an experimental metallurgist by trade, and more of an enthusiast for American ideas, was Port Pirie's acting general manager in late 1918.[39]

Interest in overseas developments, and a number of specific problems at Port Pirie, drove the BHAS reform program. There was a belief that better facilities and more company-provided services would lead to greater efficiency at the plant, create a more co-operative environment for labour relations, and make the company a more attractive employer, thus securing a more stable workforce. For BHAS management there were two other factors: the BHP period at Port Pirie, where inefficiency appeared to go hand in hand with distrust of management, poor working conditions and labour tension; and industrial relations at Broken Hill, where an active and militant labour movement foiled management initiatives at almost every turn. The principal message that BHAS management took from all this was that a 'moderate' approach to industrial relations, plus better facilities at the workplace – an approach sometimes known as 'welfarism' – could improve the efficiency of the smelter, and insure against industrial trouble.[40]

Any program of reform needed to be firmly entrenched in the workplace before more ambitious and broad-ranging policies could be pursued. From 1916 to 1920 a system of workplace committees was set up to administer new benefits and services. These committees were typical of those in welfarist schemes in Great Britain, Canada and the United States, and had been brought to the attention of the BHAS board by the British Whitley reports and the Works Council movement.[41] The most important of these committees at Port Pirie was the Co-operative Council, which was formed in late 1917. The council consisted of both appointed and elected staff and workers, who were to establish and run a co-operative store to supply food, clothing and other items which were either not available at Port Pirie or were supplied at an excessive cost by local storekeepers. The council's constitution, drafted by Mussen, called for 27 representatives elected by the workforce and six appointed by the company. These 33 members would then elect four from among themselves to form the executive, with the company appointing three additional executive members. Decisions taken by the Co-operative Council or the executive could be vetoed only by the general manager. A similar arrangement was made for the Provident Fund.[42] Other committee-run schemes included the Sickness and Accident Fund, which was set up in August 1917, based on a 6d levy of all workers, and a Benefit Fund which was to help 'employees in distress or financial need' by offering loans of up to £20.[43]

By 1920 the company had made significant improvements to

working conditions and other works-based facilities. A First Aid and Ambulance Room had been established, and there were water filters and ice boxes throughout the works. Attempts had been made to install machine guards on all dangerous machinery, new change rooms were constructed, and improvements were made to ventilation near furnaces.[44] By 1920 the essential features of industrial welfarism were in place.

Remaking local society

Important parts of the BHAS strategy were targeted at the broader economy and society of Port Pirie. One of the key problems for BHAS management between 1915 and the mid-1920s was lack of skilled and reliable workers. Many men engaged on a daily basis were itinerant workers who had little or no loyalty to the company, and often left town for a job further afield or took better paid work on the wharves. The company found that between May 1916 and May 1917, 1931 new men were employed by the works, when the average size of the workforce was 1886.[45]

Management's first, and perhaps most important, means of dealing with labour supply problems (and indeed labour cost problems) was to reorganise the points of production. As Colin Fraser, joint managing director with Robinson, noted of the Zinc plant in 1916:

> men are doing work here that is elsewhere done by machines. The whole layout of the plant is bad. Work is multiplied. Mechanical transport is in most cases non-existent and material is handled and rehandled by manual labour.[46]

Throughout the late 1910s and 1920s the company's labour requirements were reduced by labour-saving technology, and by studying the efficiency of each component of the plant and 'shortening up' labour when excess was discovered.[47] By 1927–28 the average size of the workforce had decreased to 1533 (the peak was 1800 in 1918), as a result of 'increased efficiency in the plant'.[48]

The reduction in overall labour requirements, however, was a long-term strategy; it did not solve the daily need for labour, particularly during labour-intensive construction work. The company used agents or ads in the daily papers in Adelaide and Melbourne during periods of severe shortage. Workers with the requisite skills and experience were sometimes given free transport to Port Pirie. Some of the more highly valued tradesmen were also offered financial assistance to bring their families with them.[49] The broader management agenda, then, aimed to attract more workers to Port Pirie, and to encourage a more settled community of married workers, thus reducing the company's reliance on single itinerant men.

Public facilities and 'respectable' recreation

BHAS went to considerable lengths to provide employees with 'respectable' forms of leisure and recreation. In 1919 the Weeroona Holiday Camp was established, about 29km across Spencer Gulf. A steamer was provided free of charge to take employees and their families to the camp over the Christmas/New Year period. The camp was later moved to Port Flinders, 16km north of Port Pirie – this area is still commonly, though unofficially, called 'Weeroona'. It was another attempt to gain a more settled and reliable workforce, and highlights the influence of the state of the local labour market on labour management strategies.[50]

The camp had a central kitchen and dining room, and was run by a superintendent. It provided opportunities for staff, workers and their families to mix in a casual atmosphere. Eligibility to attend the camp was linked to achieving a bonus of two weeks' leave for employees who worked the requisite number of days in the year. Plans for the camp were undermined, however, by the closure of the smelter from August 1919 to May 1920 (a result of the Broken Hill strike). The camp also suffered a blow-out in costs in 1919, and by mid-1922 it was being used solely as a convalescent home for 'leaded' workers from Port Pirie.[51]

The most significant company recreation project was the construction of the Soldiers' Memorial Park and Children's Playground, on the site of the old 'Central Parkland'. The idea for the new park or garden was first put forward by Gerald Mussen at a council meeting in February 1917.[52] Outlining his goal as the pursuit of 'industrial peace', Mussen suggested a new park and general town improvements.[53] In March 1917 the company offered to help with any plans to beautify the town.[54] In June, at a meeting held 'in connection with the inauguration of the parklands improvement scheme', Mayor J.S. Geddes argued that the town needed 'some suitable spot on which later they could erect a monument to those who had given their lives for the Empire'.[55] By mid-1919 BHAS had spent £5769 on materials and labour for the new park.[56] There was also a company-sponsored working bee in August 1918, attended by many workers and townspeople.

The park included a bewildering array of facilities and equipment: croquet lawn, bowling green, tennis court, quoit ground, basketball field, running track, wading pool and gymnastic equipment. These were the activities of the respectable middle class; the working-class youth of Port Pirie would probably have preferred an Australian Rules or soccer ground.

BHAS management also constructed an outdoor theatre in the new park – it was Robinson who saw the political value of cinema. A Melbourne-based film maker, F.A. Jeffrey, was hired to film the working bee, and a picture pavilion was erected in the playground for the first anniversary of the working bee, in August 1919.[57] Plans drawn up for the Weeroona 'Health Resort' in 1918 also included an 'open air cinema'. Robinson's views clearly had some influence. Ultimately, the operation of the outdoor cinema was thwarted by council regulations. However, the cinema plan underlines the fact that the Soldiers' Memorial Park aimed to encourage the values, attitudes and practices deemed appropriate by the company. The workers and their families could enjoy healthy outdoor sports during the day and 'clean elevating pictures' at night.

These activities were designed to encourage friendly, social contact between staff, workers and their families. This was also true of the company picnic, first introduced by BHP in 1904 and taken up with greater vigour by BHAS in 1917.[58] The company picnic was commonly held at the showground at Crystal Brook, a rural town some 30km from Port Pirie. A picnic committee organised a year-long series of meetings and fundraising events leading up to the big day in early October. All workers were given the day off, which was an effective way to encourage participation. The picnic committee was formalised in 1927, as the BHAS Picnic and Sports Committee, with the company offering matching funding for any money it raised.[59] From BHAS's point of view, picnic day, which included a lunch, games and races, was an expression of the shared interests of management and workers; the activities for women and children signified BHAS's interest in the employee beyond the workplace.

There was a perception among senior managers that married men were more reliable than single men. In 1917 General Manager William Robertson complained that the men being sent by the labour agent in Adelaide were 'a lot of absolute "wasters" who only work for a few days then get drunk'.[60] Single men were seen to be prone to alcoholism, less likely to become permanent members of the BHAS workforce – and many were not of British origin. In 1915 Robinson complained that labour requirements at Port Pirie could only be met by employing 'foreigners' – Italians, Greeks, Maltese, Swedes, Dutchmen and Danes – who were 'mostly deserters from ships'.[61] In the following year, out of a total workforce of 1875 there were 662 single men.[62] Such workers were slowly forced out of employment at the smelter, especially after the Royal Commission into lead poisoning, whose findings, according to the company's interpretation, justified a hiring policy which placed married 'Britishers' first and single foreigners last. The policy for firing workers also favoured married British workers: single foreigners were the first to be retrenched. The effectiveness of these policies can be gauged by the reduction in the numbers of single foreign workers at the smelters. There were 651 'foreign' workers in March 1925, but only 189 by June 1928.[63] Many of these men returned to their countries of origin, but some found work locally, in fishing and farming in particular.

The proportion of single to married men at Port Pirie was also related to housing. This was another area of the company's interest. BHAS conducted a wide-ranging investigation into housing availability and quality at Port Pirie in 1916 and 1917, partly because they knew that under BHP management the labour supply problems had been connected to the housing shortage. The shortage in the town dated back to the 1890s, when employment had expanded rapidly.[64]

The housing shortage had a very real impact on the company's labour supply. In 1918 Robertson reported that 'several of our fitters left owing to their inability to secure accommodation for their families'. He also noted that 'our new foreman rigger ... threatens to leave unless a house can be obtained for him very shortly'.[65] However, BHAS action in the area of housing was modest. The company leased a boarding house – 'Olympia' – in April 1916 which accommodated 100 single men, and in mid-1918 four concrete homes were constructed for workmen. The company relied on the private housing market to provide for its employees, and further action in this area was precluded by the shutdown of the smelter between August 1919 and May 1920.

The company introduced yet more measures designed to encourage married men to stay in Port Pirie. In 1917 a Co-operative Store was established 'with the express object of attacking the high cost of living'.[66] The store initially only sold boots, clothing and tobacco, to an almost exclusively male clientele. It soon expanded into groceries, dairy produce, and drapery – areas where women were the key consumers. The company also began selling firewood, ice and coal at cost price to employees in August 1917. BHAS were not solely motivated by benevolence: reducing the cost of living also reduced pressure for wage increases. After the introduction of the Industrial Code in South Australia in 1920, wages were determined by the Board of Trade, using the living wage index.[67] The board set wage rates for particular regions based on the Commonwealth Statistician's index of prices and other evidence. By reducing the cost of living, the company could reduce their employees' wages. The Co-operative Store, and the cost price firewood, coal and kerosene, were also strategies to make the town more appealing.

Labour reaction

The BHAS agenda found cautious acceptance among the majority of Port Pirie unionists. The company's emphasis on discussion and dialogue dovetailed with the AWU's preference for arbitration and cooperation. It was also a welcome change from the confrontation of the BHP days.[68] One of the first welfarist schemes, the Sickness and Accident Fund, was approved by a majority of the local AWU membership – 315 to 198 – despite a directive from local union officials to vote against it.[69] By late 1919, 75 per cent of the workforce were members of the Sickness and Accident Fund.[70] There was still resistance, however, from a number of local unions, including the influential WWF. The editor of the *Port Pirie Advertiser* noted that some unions were against the reforms 'on the principle that there must be some ulterior motive in the company's mind'.[71]

One area that did provoke controversy was the company's plan for a dental clinic and free yearly check-ups for employees, first suggested in 1923. At a ballot organised by the AWU on the yearly medical examinations, 384 members voted against and 83 voted for them. Opposition was led by AWU organiser and Port Pirie councillor William Robinette.[72] Even the more moderate unions, such as FEDFA, moved that a government inspector be appointed so that truly independent medical advice could be provided.[73] A handbill issued by Robinette noted that:

> The first symptoms of lead poisoning are shown in the mouth and by submitting yourself to dental examination on the Works, you may give the Company knowledge of your condition and possibly bring about your dismissal in a surrepticious [sic] manner without receiving any compensation.[74]

Robinette's campaign culminated in the 1925 Royal Commission into lead poisoning, and in further campaigns against company-provided medical and dental services in 1926. Though evidence given before the Commission tended to vindicate Robinette's fears, by March 1928 the Dental Clinic had 778 members and there had been 1181 visits to it in the past year.[75] In other areas the company was even more successful: by May 1924 around 1000 workers out of a workforce of 1600 were members of the Sickness and Accident Fund.[76]

Opposition to the company's broader cultural and political strategies was virtually non-existent. Workers enthusiastically embraced the Children's Park, with 1000 contributing their labour free to the cause. In July 1918 Mussen wrote to Fraser about how skilled workers had participated: 'They all want to make something for the playground that will be a credit to their craft.' The blacksmiths, for example, had made 'complete legs and backs for garden seats every five minutes. It was a remarkable piece of organisation and skill.'[77]

The Co-operative Store was another successful venture. In 1920 the store moved to larger premises because of its popularity.[78] Total sales for the financial year ending June 1922, for example, were £44,875, and the store recorded a profit of £820.[79] In the following year the traders of Port Pirie expressed concern that it was taking away their business, particularly when they learnt it was expanding into new areas, such as drapery.[80] Likewise, the Holiday Camp attracted 500 campers in its first year of operation, along with 1200 visitors from Port Pirie.[81]

Nevertheless, underlying fears about the true motives behind these strategies remained. The concerns of some union officials about the legitimacy of the Co-operative Council had some basis in fact, though they had no evidence at the time. Management presented highly abridged accounts to the Executive and the General Council in an effort to hide the large profits the store was making. In December 1925 the General Superintendent, W.A. Woodward, reported to the Melbourne office that a 'secret reserve' fund had £1405 in it, though the printed accounts listed a profit of only £204.[82] These inadequate accounts produced criticism from union representatives; Harold Summerton, secretary of the local branch of the ASE, stormed out of one meeting after his questions on the accounts were ruled out of order.[83] The company's hiding of the real profit was revealed in a candid letter from Superintendent Woodward in 1936. Noting the requests for wage increases by workers and staff, he suggested that a profit of no more than £201/6/7 be disclosed.[84] The company presented abridged accounts to the Co-operative Council until the mid-1930s.

The absence of significant formal political or industrial opposition should not be construed as evidence of total support for the company's program.[85] There is evidence of cultural resistance to company-sponsored recreation and leisure. Though workers had shown great enthusiasm in helping to construct the park, they were less interested in using its facilities once it was finished. In 1921 Mayor J.F. Jenkins lamented that parents had shown a lack of interest in the park.[86] In 1922 he took up this argument again:

> young children continue to make good use of the Children's Playground, but there is very little evidence of any appreciation by the parents of the facilities provided, at great cost to their Council, for the welfare of their little ones.[87]

Similarly, the BHAS picnic was a company-sponsored version of older labour movement rituals, and such rituals did not simply disappear. The local labour movement had its own picnic rituals, often associated with Labour Day. They included sports, speeches, a lunch, and a concert in the evening.[88] The Port Pirie waterside workers held their annual picnic at Crystal Brook showground, with foot racing, other sports and games, and a dance in the evening. An advertisement for the event called upon waterside workers to 'Make this a Picnic for your families', and noted that 'Children will be specially catered for by a Special Committee' – local unions also understood the importance of family-based recreation in winning loyalty.[89] In the 1920s the WWF donated £75 annually to the union picnic.[90] It is significant that the

BHAS annual picnic was scheduled for early October, when workers in South Australia and elsewhere celebrated Labour Day. The timing suggests a campaign to undermine the popularity and influence of the labour celebration.[91] The two types of picnics are indicative of a clash over the control and organisation of public forms of recreation, and highlight the way in which the industrial struggle between labour and capital spilled over onto a broader cultural/political canvas. The form – and even the venue – of the company picnic was strikingly similar to both the Labour Day picnic and the WWF picnic.

The picnic also highlights the role of the WWF in resisting the BHAS agenda. More than any union at Port Pirie, the WWF developed its own industrial and cultural agenda beyond the company's influence. Along with their Waterside Workers Picnic, the WWF also ran a Bakery and Co-operative Store in Port Pirie from the early 1920s.[92] The union was active in many of the non-workplace areas that BHAS wished to influence.

The broad management strategy of making the local economic and social infrastructure more appealing to workers gave the company considerable leverage in determining the extent and the quality of its labour supply; by the late 1920s the local newspapers were extolling the virtues of 'industrial co-operation', and vast numbers of workers and their wives shopped at the company store and attended company-sponsored forms of recreation. The twin objectives of reforming the workplace and reforming the community of workers were closely linked in the minds of the directors and managers of this major Australian company.

Making a town – growth after 1945

After World War II, BHAS maintained its dominant economic position in the town's labour market. By now, its continued operation was a necessity for the town's prosperity. Many oral history respondents noted the generational line that ran through employment at BHAS – sons were expected to follow fathers and grandfathers to the smelter.

This was formalised in labour recruitment: sons of smelter employees received preference until 1967. This was similar to, though not as comprehensive as, the Barrier Industrial Council's local preference agreement.[93]

By 1947 the town's population had expanded to 12,019 with a slight dominance of males over females (6160 to 5859). There were 329 males and 187 females born in the United Kingdom. There was modest cultural diversity, but many workers from Europe, in particular Italy, Greece and the Scandinavian countries, soon moved on. In 1947 the Europe-born population was 1003, a little over 8 per cent of the total population. There were 21 Asian-born residents, mostly working as market gardeners, or in the town's laundries.

The port was an important factor in the local demography. Let us consider these effects from the early 20th century and then pick up the thread of the post-1945 period. In port communities, a land-based and a maritime labour market met. The presence of a seafaring and waterfront workforce, with large groups of workers from many different places, meant that port communities were always exposed to labour market influences. In the 1890s, for example, an estimated 2000 sailors a year moved through Port Pirie, staying from a few days to many weeks, depending on the loading and weather conditions, and on labour need and supply.[94] All this related to harbour improvements funded by the colonial (and later state) government. The government dredging (from 1874), the embankment works on Ellen Street (from 1875), and later government-built wharves, slowly forged a successful commercial harbour.[95]

This maritime labour market was a unique element of Port Pirie, making it different from the other towns covered in this study. Itinerant workers gave Port Pirie a broader cultural mix and sense of cosmopolitanism, but it also led to local 'white' workers setting up defensive coalitions against 'outsiders'. Port Pirie had a number of brothels and prostitutes, partly to cater for the shipping workforce. Off-the-record comments from Port Pirie residents suggest that the back of the Barrier Hotel in Ellen Street, and at least one establishment near the

International Hotel, were well-known sites.[96] The presence of large numbers of sailors in town gave the Port Pirie hotels a potential for unruly behaviour through groups of men who were charged with alcohol and lacking the familial and workforce links of local men.

The port also enabled the migration of a distinctive community: Italian fisherman and their families, most of them from the port town of Molfetta in the province of Bari. Molfetta had a long tradition of workers moving between Mediterranean ports for a period and sending money home. These were not single journeys of migration but multiple trips with regular returns. Vito Caputo, believed to be one of the first Italian migrants, arrived in Port Pirie in 1885, recognised the potential of the area for fishing, and began a long line of family and community migration from Molfetta. Port Pirie appeared to be an ideal place to establish a fishing industry, as there were abundant fish stocks and few local competitors. By 1900 there were 45 fisherman at Port Pirie, around half being of Italian origin. The 1901 Census records 18 Italian-born males and 4 Italian-born females in Port Pirie. By 1921 the Census records 44 males and 18 females born in Italy. Italian migration to Port Pirie, and to Australia generally, increased in the 1920s, in response to worsening poverty in the rural areas of Italy and tighter immigration controls in the United States. Italians were also common in the smelter workforce by the late 1920s.[97]

There was a mix of compromise and conflict between the Italian and Anglo-Celtic communities, particularly before 1945. Disputes over fishing resources and fishing practices led to some complaints. The outbreak and upheaval of war, conscription and returning soldiers also saw some conflict break out in Ellen Street. An incident in 1914 is indicative of some of these ethnic tensions. Three Italian men were standing in Ellen Street when they were approached by three Anglo-Celtic men, who asked them for money. When they refused, the Italians were called racist names and a fight ensued. The police arrived and arrested the Italians. Eyewitnesses indicate that a crowd had gathered, and continued to punch and kick the men, suggesting a more general hostility towards Italians.[98] The Italians were often assumed to be anarchists and radical militants, and this may have further inflamed local opinion around this time.

The fight between the Italian and Anglo-Celtic men was not an isolated case. Racist views were common in the period, leading to more sustained racial violence in Broken Hill, Queenstown, Port Adelaide and Kalgoorlie in the 1920s.[99] They were also seen in the 1925 Royal Commission into Lead Poisoning: the commission concluded that the high incidence of lead poisoning at Port Pirie was not the result of workplace conditions but the outcome of the poor living conditions and indolent habits among foreign workers. Thus migrant workers could be blamed and further marginalised. One recommendation of the commission was that foreign workers be gradually removed from the workforce; as noted above, this was indeed carried out.[100]

The Italian community was just one of many non-Anglo ethnic groups in Port Pirie before 1945. In times of economic stress and political crisis, who qualified as a local was bounded by whiteness and a particular Anglo-Celtic heritage. Some Italian residents were interned during the war. Localism was complicated by a sense of ethnic difference, but at the same time there were important points of contact between different cultural groups before 1945. This set the scene for a stronger engagement with, and acceptance of, cultural difference after 1945.

The cultural effects of the port played out in the entire community, but also had specific effects on individual residents. Bruce Steel, for example, first saw Port Pirie as a cabin boy on the *Coorabie* in 1941. Asked about the first time he saw Port Pirie, he recalled:

> One thing that stuck in my mind quite strongly from those days was the little Italian fishing boats going out. They were good. Most of them [had sails], or occasionally one with a motor in it, and he might be towing a couple of others. And then one of them had sails up with big sweeps, and that's always stuck in my mind.[101]

It had a strong impact on the young man. He returned later in life, settling in Telowie, north of Port Pirie, and then in the town itself.

Migration from countries such as Italy and Greece became more common after these countries signed agreements with the Australian government. The agreement with Italy was signed in 1951. Recession in Australia in the early 1950s slowed the wave of migrants, and immigration officers still unofficially selected on the basis of acceptable racial characteristics, so some Italian migrants were rejected because their skin was seen to be too 'dark'.[102]

In 1952 the Australian-Greek Migration Agreement was signed. In the 1954 Census Port Pirie had 273 males and 180 females born in Italy and 68 males and 43 females born in Greece. Overall, 1498 people in Port Pirie (10.5 per cent of the population) had been born in Europe. This was a slight increase on the 8 per cent recorded in 1947.[103] By 1966, 9.5 per cent of the total population of 13,964 had been born in Europe. There was only a minor change in the numbers of Italian-born people by 1966, which accords with general trends for South Australia, where the highest intake was in the early 1950s.

Migrants consolidated into religious, cultural and sporting groups, mirroring the wider consolidation of Port Pirie society. However, the dominant society at Port Pirie, particularly before 1970, was a relatively unsafe space for migrant activity. Ironically, this drove the creation of a rich and varied cultural life in the locality. For example, the Catholic church was revived through the influence of the Italians. In 1951 the Diocese was moved from Port Augusta to Port Pirie, and the new St Mark's Cathedral was blessed and dedicated in 1953 (the old church had been destroyed by fire in 1947).[104] The Greek Orthodox church also became an important centre for community activity and organising in this period. Greek Orthodox services had started in 1913 and the Church of St Paul had been consecrated in 1926.[105]

While Port Pirie was changed by these broad international and national shifts, the change was not nearly as great as in nearby Whyalla. This was because with smelter employment steady, and no new industries, there was no significant labour demand of the kind experienced in Whyalla. BHAS had significantly reorganised labour processes, moving from batch to continuous production in the late 1930s. This, plus new technology, meant that labour needs dropped considerably. In 1915 the lead refinery employed 240 men over a 24-hour period; by 1951 this had been reduced to 55.[106]

Postwar industrialisation shaped labour markets, creating significant cultural change in the steel towns of Whyalla, Port Kembla and Newcastle, but not in Port Pirie. In 1966, an extraordinary 37.5 per cent of Whyalla's population of 22,121 had been born in Europe, including large numbers of Yugoslavs, Greeks, Italians, Dutch and Spanish migrants; this was four times higher than the Port Pirie percentage.[107] The Whyalla figures were only matched by the massive growth at Mount Isa, as the next chapter will detail.

Nonetheless, postwar migrants had an impact on Port Pirie, replenishing and reinforcing existing migrant communities. New soccer clubs were set up. The Savoy Club was established by Italian migrants in 1949, the Virtus Club by Italians in 1952, and the Olympic Flame Club by Greeks in 1958. These sporting clubs joined already established Australian Rules clubs, including the BHAS Proprietary Club (obviously supported by the company), the Port Pirie Football Club and the Solomontown Football Club.

In their fierce rivalries these clubs revealed other divisions. These rivalries reflected ethnic differences, but they also reflected cultural, class and geographic differences. Solomontown was a distinct area fringing the river east of Port Pirie. It was separated from central Port Pirie by a tract of undeveloped, swampy ground between Mary Elie Street and Alpha Terrace (less obvious now due to subsequent reclamation and development). Solomontown had a large number of working-class and Italian residents, and its own small commercial area and school. Though many Solomontown residents worked for the smelters, they could not get a game with the Proprietary football team; there was a perception, at least, that they were less favoured.

Residents from Pirie West, adjacent to the smelter, were well-represented on the Proprietary team. It included a number of Greek residents. Barry Scott played ruckman for the Proprietary Colts in

1949 and 1950 alongside a rover of Greek origin. Catholics were well represented in the Solomontown team, but also gravitated to Port, since this club had its origins in the Catholic Churches League and was called St Marks from its formation in 1909 until a name change in 1941. These suburban, cultural and religious identifications existed alongside overarching town loyalty and parochialism.

The workforce figures from the 1954 Census provide a snapshot of the labour market after almost 10 years of postwar growth. There were 4199 males and 895 females recorded as 'employee (on wage or salary)', 202 males and 37 females who were 'employers', and 159 males and 51 females who were 'self-employed'.[108] Such categories are only of limited value. There was no detailed breakdown into industry groups, and the records do not deal with the informal economy, which was dominated by women, but they do show us an economy where those who worked for a wage were usually men, and at Port Pirie most of them worked at the smelter, the wharves, or for the railways. The smelter, while never completely dominant in the way that a single company was at Mount Morgan and Queenstown, was still a major employer. Even in 1977, the smelter employed 1800 workers out of a total workforce of 5976.[109] The sale of South Australian Railways to the Commonwealth in 1975, and the reduction of the railway workforce at the Port Pirie yards from 1980 increased the smelter's dominance.[110]

Wages at the smelter were good, especially for workers with a trade or skill. In 1954 Bob Krahnert was earning £54 a fortnight as a crane driver at the smelter. This was well above the basic (minimum) wage, which for Port Pirie was set at £11/2/- a week in January 1953.[111] The wage was supplemented by the lead bonus. Starting in late 1952, average wages for the men at the smelter were £18 per week plus a £5 lead bonus. The loss of the regular cost of living adjustments in 1953 and rising prices in the 1950s meant that such wages were not in fact always generous, but Port Pirie was not as affected by the rising prices as the more isolated towns of Broken Hill and Mount Isa.

Ethnicity could affect earnings in the smelter workforce. Migrants of Greek and Italian origin often worked in the low-paid, less desirable jobs. In the Bag House, for example, chief metallurgist Frank Green noted that '[t]his work is not eagerly sought and for many years it was difficult to obtain operators of British stock for the smelter baghouse, hence the crews were mainly of Greek and Italian origin'.[112]

BHAS maintained its strong welfarist tradition, with its dependence on family links between grandfather, father and son, as well as uncles and nephews, after the war. The smelter remained a workforce with large numbers of long-serving and related workers. The style of management was not to everyone's liking. Workers who had been employed in a number of different places and in different industries were particularly resistant. Bruce Steel, who had worked for the public sector, in small business and in large industrial workplaces, recalled of BHAS:

> The lords and masters, the staff men, were distinctly different from the ordinary people, even with the pursuits of things around the town. They were members of this and members of that, and ordinary workers were not at all welcome in different things like that.[113]

While Steel's view may not have been universally shared – many workers were genuinely fond of the company – BHAS staff had significantly different origins and experiences from local working-class families and constituted an identifiable elite in the town. This can be illustrated by considering the biographies of key senior staff. Frank Green, chief metallurgist from 1944 to 1955, was born in Melbourne in 1903. He was educated at Scotch College and the University of Melbourne, where he took a degree in metallurgical engineering. He started working for BHAS in 1925, and after retiring in 1955 joined the board of the company, which entailed regular trips to Melbourne.[114]

Green's story hints at another significant difference between local working-class men and the staff men; staff men were more widely travelled and their geographical orbit extended beyond the town. Trips to Broken Hill to visit the workplaces of the joint venture partners,

Broken Hill South and North Broken Hill, were common. During these trips, which were reciprocated by the Broken Hill-based staff, staff men mixed with other staff men in works inspections, social cricket matches and after hours drinking. In 1947, 50 staff men, some accompanied by their 'wives' (so named by the newspaper report), visited Broken Hill to tour the works. The wives were given a tour of the surface, and then entertained at the North Broken Hill Staff Club.[115]

Essington Lewis, who became managing director of BHP in 1921, was a staff employee at the BHP Port Pirie smelter from 1905 to 1915. He was active in the Port Pirie Quadrille Club, the Excelsior Brass Band, the Port Pirie Institute, and the Horse Show. Lewis was an excellent athlete and an accomplished Australian Rules player. He was captain/coach of the 1911 premiership-wining Proprietary team, but thereafter he played for Norwood in Adelaide when he could.[116] In 1910 Lewis had married Gladys Cowan, the daughter of a prominent local grazier and mining investor, and moved into a large house in Port Pirie South with 12 acres and a gardener. After illness in 1911, his wife moved to a home in Leabrook, in suburban Adelaide, and Lewis divided his time between Adelaide, Melbourne, Port Pirie and Whyalla until his move to Melbourne in 1915.[117]

The Somerset family moved to Port Pirie in 1917 after living and working in Mount Morgan, where Henry Somerset had been chief metallurgist. After leaving Mount Morgan in 1912 (with brief stints in Hunters Hill, Cobar and Broken Hill), the family – Henry and Jessie, and their two boys – moved to the Barrier Hotel in Port Pirie, where they stayed for three months. Being the most prestigious hotel in town, the Barrier was the accommodation choice for the staff families and a gathering point for the staff men. Then the family moved to a large home with an old garden on The Terrace, still the location of some of the town's most impressive homes. The new town was a shock to the family. The low-lying suburban flats were prone to flooding, and the damp salty soil made growing plants difficult. After sending their youngest son, Harry, to Port Pirie High School for two years, they sent him to boarding school in Adelaide. This was the fate of many middle-class children in mining and industrial towns where the local state school was seen as too rough or had a poor academic reputation. The Somerset children were close to the children of James Sim Geddes, sometime mayor of Port Pirie and timber merchant.[118] The families shared a governess for a time and attended the same boarding school in Adelaide, St Peter – incidentally, the same school Essington Lewis had attended.

Harry Somerset, who grew up in Port Pirie, soon left for an education in Adelaide and then at the University of Melbourne. He later became managing director, and member of the board, of the Australian Pulp and Paper Mill in Burnie, Tasmania.[119] By the 1950s he was spending three weeks in the exclusive Melbourne suburb of Toorak and one week in Burnie.

These vignettes show the patterns and paths of education and socialisation as strongly class based. Families of different classes came together in stores, in workplaces as employer or employee, or in larger political challenges that called for local unity, but for the most part they occupied separate social worlds.

The smelter not only shaped the labour market and the social structure of Port Pirie; it produced the very rhythms and sounds of the town. Many residents still recall the times for shift changeovers. Len Zubrinich watched the slag dumping from his back yard:

> [Y]ou could see it at night like a big blazing fire ... 24 hours a day; night, afternoon and day shift ... All of the colours of the rainbow would be in the metal.[120]

Before the days when slag could be re-processed to gain small amounts of payable material, it was readily available. Many locals believed that it was nutrient rich and encouraged plant growth. It did damp down the rising salt, but unfortunately its use meant that many backyards ended up being rich in heavy metals, including lead and cadmium.

Influences beyond the smelter were apparent, however, and

revealed the partial control that BHAS exercised and the importance of the waterfront and railways to Port Pirie society and economy. The harbour had its own rhythms, linked to the tides, the supply of Broken Hill ore, and the progress of the wheat season. Ships sounded on entry and exit, adding to the sounds of the smelter. The railways too, with engine horns sounding at crossings, punctuated the day and night, especially from the 1930s through to the 1980s, when Port Pirie was a 'break-of-gauge' station. In those times, Port Pirie was one of the busiest stations in South Australia. Nearby Solomontown was a significant beneficiary, as railway employees sought accommodation close to the yard. Before 1967, the main station was in Ellen Street, so trains would come into the very centre of the town. Thereafter, passenger trains moved to the new Commonwealth Railway passenger terminal at Mary Elie Street, half a kilometre up the road. Joy Nunan worked in the refreshment rooms at the new station: 'these times were so wonderful, we were so busy, we would scarcely be cleaned up and shelves restocked before another train arrived'.[121]

Postwar politics

Despite success in the battle over the railways in the 1870s and 1880s a strong sense developed throughout the 20th century in Port Pirie that it was the forgotten northern cousin. Battles with Whyalla, the failure to attract major new industries, and conflict with the Commonwealth government over railway infrastructure aided this sense of frustration and defensiveness.

Whyalla was eyed cautiously as an upstart town; originally based on shipping iron ore, and dependent on Port Pirie for supplies and communication, it became a major industrial centre. In 1937, when Whyalla was announced as the site of BHP's South Australian blast furnace, Adelaide's *The Mail* reported that:

> Representative opinion in Port Pirie has expressed surprise at the suddenness with which it was announced that a blast furnace for

smelting iron ore would be established at Whyalla by the Broken Hill Pty. Co., Ltd. That announcement, coming the day following the first indication that the undertaking was being considered for South Australia, gave no opportunity for interested centres to present their claims, it was stated in some quarters. Feeling that it is not too late to ascertain whether it would be possible to have the furnace here, a general meeting of trading interests has been suggested ...[122]

The location of a large industrial asset has significant flow-on effects. Once established, fixed assets are difficult to move and tend to attract follow-up investment from the private and public sectors. This was a major setback for Port Pirie — its first major failure to attract large-scale industrial investment since the 1880s.

By 1954 Whyalla's population had increased to 8598, but by 1966, it had grown to 22,121, much larger than Port Pirie's 17,789. In this period Whyalla had seen growth in shipbuilding, new blast furnaces and mills, and a major program of expansion as BHP built a new steelworks there from 1958.[123] Major industrial and urban growth north of Adelaide was championed by the Playford government, which was in power from 1938 to 1965.

It was not that new development bypassed Port Pirie altogether. A new lead refinery was complete by 1959, and a Number 2 blast furnace was operational by 1964, the same year that new wheat silos were completed. Significant wharf reconstruction had been carried out by 1965. Zinc production at the smelter began in 1967.[124] A uranium treatment plant started production in 1955, producing uranium oxide from ore mined at Radium Hill. This was heralded as Port Pirie 'entering the atomic age'. But even this most exciting of all the industries of modernity came to nothing in the long term, with the plant closing in 1962.[125] The legacy for Port Pirie is a site that requires long-term management to restrict air- and water-borne escape of radioactive materials. The site is one of the most polluted in regional South Australia, a testament to the generally low regard that state governments have held the town in. The only rehabilitation attempted was the dumping of slag from the smelter to cover the tailings and fill the dam sites. The site was finally adequately fenced in 1978, stopping local children playing in the discarded tailings and swimming in the highly toxic and radioactive dams.[126] As if this was not enough, asbestos was dumped here in 1986.

In the 1950s and 1960s successive South Australian governments were locked in a bitter battle with the Commonwealth government over the construction of a standard-gauge rail link from Broken Hill to Port Pirie. This of course had an impact on Port Pirie. The state even sought redress in the High Court in 1961. Port Pirie shared the state's frustration at again being overlooked. The conflict over railway infrastructure resurfaced in the 1980s, when the Commonwealth government took over South Australian Railways and began reducing the size of the workforce, which affected the Pirie yards and surrounding railway towns such as Jamestown and Peterborough.

The end of the postwar boom in the 1970s had a significant effect on the town. By 1975 BHAS was operating at 60 per cent capacity.[127] A sense of the times is captured by local historian Robert Donley's work *The Rise of Port Pirie*, written in the context of the 1970s' economic crisis and decline at Broken Hill:

> The majority of its residents remain fiercely loyal to the town in the face of its detractors and determined to see that it receives just treatment at the hands of those whose power gives them control over the town's future.

For Donley, 'Many of the battlegrounds are old ones and those who read the following pages may be surprised to read how often the headlines of the city's newspapers re-echo issues of the past.'[128] The population fell from 13,227 in 1971 to 11,996 in 1981.

The town's defensiveness spoke to both a general concern over the future of the local economy and an effort to protect and defend the smelter, a crucial element of the town's prosperity. Given all this, the lead exposure issue, which exploded into prominence in the early 1980s, was a vexed and controversial one for the town, especially as it occurred alongside the recessions of the 1970s and 1980s, and broader declines in manufacturing industries.

Studies of air pollution and blood lead levels were done. This was a revival of the concern about lead exposure in the workplace that had led to a Royal Commission in 1925 and subsequent company and government action. Former chief metallurgist Frank Green claimed that there were only two certified lead cases between 1953 and 1968 and none from 1968 to 1977.[129] But by this time the focus had shifted to the surrounding environs – not just of Port Pirie, but also of the other towns covered in this book. It was not the limited-duration high level of exposure at the workplace that was causing the problem – this had been moderated by improved smelter performance – but long-term, low-level community exposure (through soil or airborne presence) to lead, sulphur dioxide and heavy metals such as zinc and cadmium.

In 1976, CSIRO researchers published a wide-ranging study of the heavy metal contamination of the soil in and around Port Pirie.[130] They estimated that 20,000 to 40,000 tonnes of lead, and 250 to 450 tonnes of cadmium, had been deposited as fallout in the area; 90 per cent of this occurred prior to 1940.[131] The distribution of the highest levels of contaminations – up to 40km to the south and up to 25km to the north of the smelter – corresponded with the prevailing winds at Port Pirie, and a concentration in the hills to the east of the town related to the action of higher altitude winds, with hilly areas forming a natural barrier and deposit site. The fallout comes not only from particulate and emissions from the smelter's stacks but from later wind-blow (disturbed material being transported again). The researchers found one site, for example, where high levels of surface contamination had been ploughed into lower strata of the soil. This research was not focused on the hazards that can result from the presence of this material; that work would be done in the 1980s, following widespread concern over blood lead levels in children – at Port Pirie, and many other smelter locations worldwide.

In 1984 the South Australian government established the Port Pirie Lead Implementation Program in response to research which demonstrated that 98 per cent of Port Pirie's children had blood lead levels above 10 micrograms per decilitre.[132] Jenny Pryor was appointed as a case worker for the Lead Program in 1984 and later played a senior role in it. The initial response to this program was clear: 'The community felt very threatened ... It was the only industry and people were very protective and defensive of their industry.'[133] The association between lead exposure and lower IQ, as well as the correlation with household cleanliness and regular hand-washing, brought out this defensiveness very strongly. As Pryor noted, beyond the real issues of the physiological effects of lead, there was a sense in the community that this concern suggested that Port Pirie people were unclean, slovenly and bad parents.

Despite these challenges, hindsight shows that the government and overall community response at Port Pirie was groundbreaking. Similar programs were instituted at Broken Hill and at Cockle Creek, south of Newcastle. For a community with few resources and a heavy reliance on one industry, progress was impressive.

Port Pirie lies at the confluence of a number of transport networks, both land and maritime-based, which affect its economic and cultural makeup. The link to Broken Hill was crucial in the 1880s, and allowed the wool and wheat export town to secure the Barrier trade and its smelting industries. The town has long had close affinities with, and connections to, its hinterland, but also moments of bitter competition, initially with Port Germein and Port Adelaide, and later with Port Augusta and Whyalla. Despite – or perhaps because of – this connectedness, local society and politics have coalesced into a defensiveness that has produced a decades-long campaign to protect and enhance the town's infrastructure and its prospects. The most recent example of this defensiveness was the difficult issue of the community exposure to lead.

Nonetheless, the entire town shares a strong sense of the past, as shown by the activism and success of the National Trust from 1967.

Port Pirie

'The Friendly City' moniker adopted in the 1980s appears to have been successful, as there is a strong sense of ownership of that label and a real effort to live up to its meaning. The broad commitment to place is shared across social and cultural groups, and the historic fabric of Port Pirie is largely intact. Like Mount Morgan, history and heritage may prove to be one of the town's greatest assets. Meanwhile the large lead smelter, now owned by Nyrstar, is still in production employing over 700 and much political effort is still directed towards maintaining its presence.

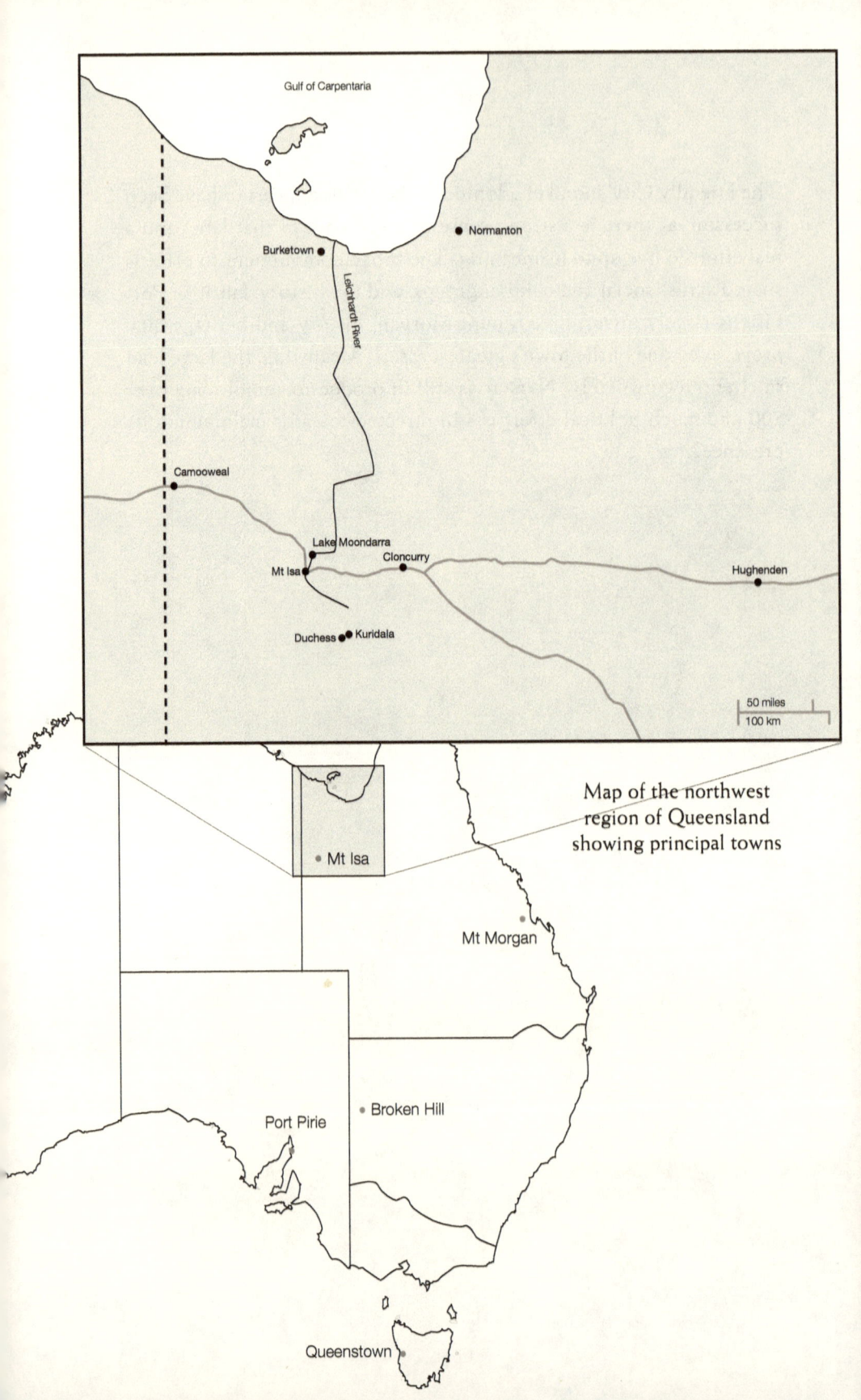

Map of the northwest region of Queensland showing principal towns

6
MOUNT ISA
Normalising outback suburbia

Mount Isa is a dry, red-dust town surrounded by the rolling hills of the Selwyn Ranges, in the northwest of Queensland. The transition from the flat, almost featureless plains to the southeast, around Longreach and Winton, to the hilly red clay country near Cloncurry and Mount Isa is striking. The country to the west, through the Northern Territory, the Barkly Tablelands, and the flat plains near Camooweal, is an equally remarkable contrast to the Mount Isa region.

In this dramatic country, with reds and browns dominating, the town centre itself is an unexpected and somewhat dissonant green, hunkered down in the lee of the mine site and the smelting works on the banks of the Leichhardt River. The town centre did not always look this way, though. During the early stage of its development, in the 1920s and 1930s, it was a straggling set of shops and businesses made of scavenged canvas, calico, iron and tin. This was an unforgiving isolated site, 1000km from the coast.

Mount Isa is one of the largest mining towns in Australia, a testament to the extent and richness of the orebody that lies underneath it. This orebody includes silver/lead/zinc-bearing ore as well as substantial copper deposits. The orebody was discovered in 1923, and large-scale mining began in 1929. A lead smelter followed in 1931. By 1960 Mount Isa produced 60 per cent of Queensland's mineral wealth, and employed over 50 per cent of the state's mining workforce.[1] The mean maximum temperature is 37.7°C in December and 37.6°C in January. There are, on average, around 28 days per year where the temperature goes over 40°C and 138 days where it goes over 35°C. Winter nights can be cool to cold, especially since the town lies at an elevation of around 350 metres. The annual rainfall is 386.4mm. Most of the rain falls in summer as monsoonal weather systems move into northern Australia.[2]

A central part of the modernist narrative was the transformation of harsh and difficult landscapes into productive industrial and urban environments. This chapter explores this theme with respect to Mount Isa. In Newcastle, much was made of the new (in 1915) BHP iron and steelworks being built on the site of an unproductive 'swamp'. The large anchoring pillars of these mills provided reassuring solidity. In Queenstown, an area of sometimes wild and difficult weather, and the creation of a town environment with palls of smoke rising from industry and homes was satisfying to locals and visitors alike. But in Mount Isa, the extent of this challenge appeared the greatest; it was sometimes presented as the 'taming of the tropics'. Mount Isa exemplified the general concern or anxiety about the fate of the white race in the tropical regions: were white people capable of living and working in this climate? This was a continuing theme, and added to the challenges of isolation and distance.

Around the margins of this framework lie the experiences and memories of locals, which both reaffirm and offer new perspectives on the modernist narrative. Locals were part of the ferment and excitement of industrial modernity, yet they also sought refuge from its more dehumanising elements in the fellowship of others, and contested it through industrial action.

The beginnings of a mine

In 1923 John Campbell Miles was prospecting along the Leichhardt River when he happened upon heavy and darkly coloured rock specimens. These piqued his interest, so he sent samples to the government assayer at Cloncurry. Miles continued to investigate the area along with others – some already there and some who arrived in the following 12 months. Prospecting and mine development work continued into early 1924. The early reports were very positive; it was clear that Mount Isa was a significant find. Most of the smaller leases pegged in this period were eventually sold to larger and more heavily capitalised companies. Mount Isa Mines Ltd (MIM) emerged as the major mining company in the area by 1925, though some dogged competitors persisted.[3]

The Kalkadoon people were witness to these and earlier developments. There were a series of violent confrontations, known to local Indigenous people as the Wild Time, including the Battle Mountain massacre in 1884.[4] By the early 20th century, though they were subject to increasing state intervention and control, they had found a place in the rural labour market on pastoral stations. They were still very much present in the area in the 1920s and 1930s, often guiding prospectors or saving them when they became lost or injured.[5]

By 1930, there were 1500 workers on the field.[6] In 1943 copper started being worked in commercial quantities, in response to the Commonwealth government's war-time demands. The presence of a substantial copper deposit, entangled with the silver/zinc/lead orebody, makes Mount Isa unique in geological terms, and gave the operation a broader base.

The northwest

There was a highly specific regional context for the development of Mount Isa, and it requires elaboration. When the orebody was confirmed in 1923, the challenges of the mining field and town development seemed almost insurmountable. There was a difficult semi-arid desert environment to manage, and the distances involved were enormous: the Gulf of Carpentaria lies 400km to the north, and the Queensland coast, via the road to Townsville, is 900km away. Even the isolation of the west coast of Tasmania was trumped here.

Local and regional histories have focused on the struggle of white settlers. Mining towns commonly have a pioneering ethos that celebrates a select group of 'founders', usually white men who 'discovered' the field or helped establish mine production. This is the role occupied by John Campbell Miles. In many texts, and in the public history of the town, the settlers' struggle with the environment, rather than their equally dramatic and far-reaching struggle with the Indigenous people of the area, is presented as the central story. However, the late 19th century saw considerable Indigenous resistance, some passive and some violent; only the Battle Mountain massacre has found a place in local histories.

Despite the idea that Mount Isa was a town hewn from a vast and unpopulated outback, there were settlements in the region before 1923. There was much interest in gold and copper prospecting throughout the region. Camooweal, a rural town on the edge of western Queensland (180km from Mount Isa), was proclaimed in 1884, and developed as a pastoral service town. It drew its supplies from Normanton, near the Gulf – this pattern of connection and development pre-dated Mount Isa. Now a fuel stop for many before crossing the Barkly Tableland, Camooweal has a population of only 300.[7]

Cloncurry is another town that pre-dated Mount Isa and was an important stepping stone towards the discovery of the Isa orebody. This mining town, about 110km east of Mount Isa, developed from the 1860s and had cycles of boom and bust related to its gold and copper deposits. Cloncurry was an important transport hub for the railway from Townsville, which was completed in 1908. A rich copper mine at Duchess – 100km south of the future Mount Isa site – developed from the early 1900s. In 1925 the Queensland government passed the *Duchess to Mount Isa Railway Act*, and that line was completed in 1929.

So the new town of Mount Isa had links to the smaller regional centres from its early days. Cloncurry was a vital supply depot and staging post. The Shire Council at Cloncurry was the body first responsible for the development of Mount Isa. As happens in many such cases, those in the new town protested against 'outside' control. Cloncurry Shire Council, for Mount Isa residents, was a distant authority with few resources and little incentive to focus on them. It became a target for dissatisfaction, uniting disparate elements of the community.

Population growth at Mount Isa was a contrast to developments elsewhere in the northwest. Some towns had virtually ceased to exist, as in the famous case of Burketown near the Gulf of Carpentaria, which began with a flourish in 1865 but from 1867 'was virtually abandoned', ravaged by tropical disease and isolation. These were important comparisons for those trumpeting Mount Isa's development. While other towns struggled, by 1931 there were already 6000 people in Mount Isa. The town is part of the 'North Western Division', which in 1959 included only three towns with populations larger than 1000; Mount Isa (7500), Cloncurry (11,280) and Hughenden (1780).[8] With a dark cloud hanging over Queensland's other major base metals mine – Mount Morgan – in the late 1920s, the possibility of successful development at Mount Isa was of keen interest to investors, the state government, and mining workers.

One key external influence was the state government. It intervened in some areas, leaving others to MIM. The mine operation was sustained by major investment in transport infrastructure: the Cloncurry–Mount Isa railway extension, which was completed in 1932. Another, more controversial, issue was the lack of government commitment to workers' health and safety, most clearly expressed through their lack of response to the sharp rise in lead poisoning cases once the lead smelters were brought into production in 1931. Crucial services such as local inspectors and mines department offices were not established until after 1950. Similarly, in the provision of health services and planning for a new town, modest government provision or inaction helped create the void for MIM to step into. State government policy, as much as MIM intention, helped create the company town elements of Mount Isa.

Making a town

Large parts of the new town actually came from declining towns in the region. Historian Noreen Kirkman interviewed local resident Vera Olsen in 1983: 'It was quite a township there [in the late 1920s].

Everything seems to have come from Kuridala.'⁹ Kuridala, some 125km southwest of Mount Isa, developed as a copper mining town in the late 1890s, but its decline and eventual collapse coincided with the Mount Isa discovery. And once the Duchess–Mount Isa railway line was completed, in April 1929, scores of workers, some with their families, moved from Duchess to Mount Isa. Mary Board, who travelled through the area in December 1929, found Duchess a 'depressing township', with both the mine and the smelter long closed.[10] From Kuridala, Duchess and further afield, the new arrivals strained local infrastructure. Decent housing, a reliable water supply, schools and a hospital were all urgent issues.

This was a difficult stage for the town. Establishing a new town using buildings from failed nearby townships was common in mining towns: the Duchess Catholic church was relocated to Mount Isa in 1929 and its Methodist Hall arrived in 1931. As noted, many of those early residents were from the same places as these displaced buildings, and that may well have added to their sense of dislocation. The new town site was officially selected by the Cloncurry mining warden in June 1924: 'on the east bank of the Leichardt River, north of the main crossing'.[11] Despite the difficulties, the speed and sheer energy of town making is apparent. Even by 1924, within one year of Miles' discovery, two open-air cinemas had been established.[12]

The Mount Isa Progress Association had also been formed.[13] The committee was a mix of professional and commercial men and senior staff from MIM, who ensured that company interests were served by the committee. One of its first chairmen was Harry Smith, proprietor of one of the open-air cinemas and owner of Smith's Hall. That such an association should form at Mount Isa in this period is no great surprise – the roads, the water supply, and telegraphic communication were all issues of considerable concern. Also, the surrounding towns of Kuridala, Duchess and Cloncurry had all had progress associations.[14]

Appropriate accommodation was very difficult to find in this early period. Kirkman interviewed Melba Saslavski (née Evans) in 1984. She recalled her original accommodation:

> It was a corrugated iron building; it was divided off. We had one end, and the other end belonged to the [shop's] owner, George Milthorpe. We slept in there. Then the cooking and eating arrangements were off [to the side of] the potato and onion stores ...[15]

As one visiting journalist noted in May 1924, 'The field itself is in its swaddling clothes. It is the era of hessian and canvas tents, and wood and iron buildings at Mount Isa to-day.'[16]

There was no uniform approach to town building. Some efforts were sectional, in the interests of the mine workers or the storekeepers. Australian Workers' Union (AWU) members worked as miners and on the surface from 1924, and an AWU organiser from Townsville was a regular visitor from 1929. The first major industrial action occurred in 1932 and 1933, after a disastrous increase in lead poisoning among the workers at the lead smelter.[17] The storekeepers had shared interests with MIM, but with the company active in so many areas of the local economy – including running a co-operative store and providing ice, firewood and water – they were also competitors. This uneasy relationship highlighted the unfinished nature of the MIM company town project: company control was extensive but not complete.

The small mine owners and operators, known locally as 'the gougers', were also a strong force from 1923. The Labor Party, which was in power in the state at this time, had an ideological commitment to small mines and small owners, so the state government encouraged small operations as long as they could be shown to be productive and profitable. It also regulated the maximum size of a mining lease and put conditions on the sale of leases to larger companies.[18] Nonetheless, there was an inexorable move towards lease consolidation. The challenges of Mount Isa were such that regional capital would never be sufficient; only major UK and US firms had access to the huge amounts of capital required. By 1930 MIM had completed an effective consolidation of the field into a one-company operation, initially with finance from Leslie Urquhart's Russo-Asiatic Consolidated Ltd, and later with

American Smelting and Refining Company Ltd (ASARCO) backing. After that the gougers were a declining force.

Joining and complementing the public organisations of men was the Country Women's Association (CWA). Its work dealt with both the domestic world of family and children and the public world of meetings and formal organisation. In May 1928 some 20 women gathered for the first meeting of the Mount Isa branch of the CWA. By 1931 the branch had raised sufficient funds to complete a CWA Hall. It focused on providing a woman- and child-friendly space in town (the town's hotels and clubs were overwhelmingly male spaces), assisting expecting or new mothers, and providing an established network for members arriving from other branches.[19]

The Progress Association and the CWA are examples of the key role played by town-based civic organisations. The churches played a similar role, establishing their congregations very early, and providing a space for worship, fellowship and social activities. In the sporting area, clubs formed quickly. The Mount Isa Rugby League Association was active from at least 1929, with clubs including Brothers, Wanderers, Railways, The Isa Mines, Black Star and Town.

This was the beginning of a long tradition. The Isa is a rugby league town, unlike Queenstown, Broken Hill, Port Pirie and Kambalda, where Australian Rules dominates. From the early 1930s the Mount Isa representative team regularly travelled to nearby regions for end-of-season matches. These games were anticipated with relish. In 1929 the Mount Isa correspondent for the *Townsville Daily Bulletin* claimed, 'We are of the opinion that we will defeat any town in the North of Queensland.'[20] Isolation made arranging such games difficult, and exacerbated an at times tense relationship between Mount Isa and the North Queensland Rugby League Association, which was usually dominated by men from Townsville and Cairns.[21] This town-based competitiveness around a sport that Mount Isans came to see as their specialty added greatly to a sense of local identity.

Within a few years, the elements of a viable and networked community across the domains of civic, religious and sporting life were established. These activities were pursued with great energy, perhaps hoping to tame what appeared to most white newcomers as a wild and unfamiliar environment.

External influences

As a new town with considerable movement of goods, equipment, mined ore and people in and out, Mount Isa was very exposed to outside influences. The connections to the northwest region have been covered, but there were other pathways. Mining towns often promoted and maintained links to other mining and industrial towns, with the hinterland highways between them becoming conduits for information, labour and technologies. E.T. 'Taffy' Badham, whom we have already met as a FEDFA union official at Mount Morgan in the 1940s, spent three years at Mount Isa – 1934 to 1937 – before moving to Mount Morgan. He was offered the Mount Morgan job by Isa manager A.F. McAskill, the chief mechanical engineer at Isa, who had moved to Mount Morgan. Badham noted that some union organisers at Mount Morgan had come from Mount Isa.[22] Large numbers of Mount Morgan people moved to Mount Isa in 1929: their mine had closed and Mount Isa was starting up, with work available at the mine, on the new smelters and on the railway line.[23]

Badham's story is one of many linking Mount Isa with other mining and industrial towns. Once again, Broken Hill looms large. The Mount Isa orebody was constantly compared with Broken Hill.[24] The *Barrier Miner* included reports and updates from the northwest Queensland field from the beginning.[25] There was regular movement of engineering experts, senior personnel and workers between fields, as they were two of the largest mining operations in the country. Many senior Mount Isa managers, such as W.H. Corbould and George R. Fisher, had experience at Broken Hill. Trade unionists too were aware of developments in other towns.

This outside exposure reduced as Mount Isa became more clearly and firmly established. In one area, however, external forces would always dominate. All mining and industrial towns felt that their fate was ultimately decided by distant boardrooms, directors and financiers. MIM's board of directors was originally in Sydney but it moved to London, then New York, then Brisbane by 1933. Control by distant outsiders added to towns' sense of identity, which was built on the common experience of the struggling wage earner.

Life in a company town

The speed of development at Mount Isa in the 1920s and 1930s is a distinctive feature of this outline so far. This can be explained by the amount of capital behind the operation and the commitment of the Queensland government to the success of Mount Isa. By 1930 MIM was financed and run by British-based Mining Trust and a US company, ASARCO. MIM had invested some £3.7 million by mid-1931.[26] The challenge for the company was to make Mount Isa a desirable place to live. This was the key to a good labour supply. Again, early managers, such as Julius Kruttschnitt, believed firmly in the value of married workers. Mount Isa would have to be 'normalised' for family life so as to attract labour.[27]

This policy developed in two distinct phases. The initial phase involved building rudimentary workers' housing, a dam, the company hospital and the company store.[28] This all came after uncertainty regarding the town site. In an area that later became known as Mineside, the company designed a central area of a club, shops and a school, surrounded by curving streets with detached houses, and barracks for staff and single men. The initial plan included 263 houses and 18 staff houses, barracks, and sites for 60 tents. This was all completed by early 1931.[29]

By 1939 there were 312 company houses on the mining lease, plus a company-run general store, a butcher and an iceworks. Some of Mineside's company houses had well-tended gardens.[30] These gardens were a powerful statement of the ability of the company to transform the environment. MIM was changing the topography through mining, and the environment through the provision of a secure water supply.[31]

Other divisions in the local society developed. Among the waged workers, skilled tradesmen were paid more, and had access to company-provided housing. Unskilled workers were left to fend for themselves – in the town's boarding houses or in tents and humpies along the river.

The division between wages and staff workers applied here as it did in Mount Morgan, Broken Hill, Queenstown and Port Pirie, and formed a rough dividing line between the working class and the middle class. Wages workers – unskilled men and men with trades – were employed on a daily, sometimes hourly basis, while the staff workforce, with a technical or professional qualification, were on monthly contracts. Wages employees were usually required to be a part of a union while staff were usually required not to be.

At Mount Isa the staff component of the workforce was a small, close-knit group. Many of the early staff arrivals were Americans from the ASARCO parent company. The US senior staff included chief geologist Roland Blanchard, smelter specialists M.J. and M.R. Carrow, mine superintendent Frank Kane, and mining engineer and smelter designer Charles Mitke. These men were working with US-designed plant, including Mitke's ill-fated lead smelter, which required significant reconstruction and rebuilding to smelt the complex Isa ore.[32]

The US staff also included the new general manager himself, Julius Kruttschnitt, who came from an ASARCO operation in Tucson, Arizona. Under the new financing agreement of 1930, ASARCO took over management from The Mining Trust. Kruttschnitt, or J.K. as he was widely known, arrived just before Christmas 1930 with his wife Marie, and was manager until 1953. In a remarkable series of reminiscences given to Kruttschnitt's biographers, long-serving staff described the vibrant social world of the staff, full of dinners, balls, celebrations and fundraising events. Staff spouses (all women at this point) were part of the MIM family and were expected to be socially active and useful. Helen Browne, wife of the chief draughtsman, noted that Marie Kruttschnitt:

> did for the women and our life up there the equivalent of what J.K did in the field. She was always beautifully gowned and set a high standard in the entertainment area.[33]

Marie Kruttschnitt was the daughter of a San Francisco real estate agent. She married Julius in 1907. She and her family were commonly mentioned in the social pages in San Francisco before she married and later in Tucson. When she married, *The San Francisco Call* described her as a woman of 'exceptional beauty', and noted that she had a 'reputation as a writer of short stories and is gifted in music and as a linguist'.[34] Marie Kruttschnitt's activities at Mount Isa extended beyond MIM and work-based social events to local volunteer and charity organisations including the CWA, the War Comfort Fund, and the Mount Isa Bowling Club. In one rare interview, when asked what she thought of Australia, she reportedly said:

> Now you are in a hurry, so am I. So let's come straight to the point: For your own good why don't you talk more about yourselves? You honestly have a lot to brag about and you have heard that opportunity don't like standing on the door-step and what happened to the man with ten talents ... I'm leaving for America for a short holiday and a glimpse of the two children I left over there. But you may be sure I have already bought my return ticket and I shall be mighty glad to return to Australia in three months.[35]

She died in September 1940, in her early fifties.

Making a town – after 1945

Both lead and copper prices increased steadily after 1945. Lead peaked on the London Metal Exchange at over £160 per ton in 1951, coming off regulated prices as low as £15 during the war. Copper had stronger price rises than lead, moving from £80 per ton in 1947 to a peak of £347 per ton in 1955.[36] The strong copper prices underpinned MIM's profitability after 1947, with lead, silver and copper in full production from 1953. In 1952 MIM announced a record net profit of £2,179,758, dwarfing the returns from Mount Morgan and Mount Lyell.[37]

Strong growth and profitability meant that MIM's economic reach extended well beyond the immediate region. In 1959 a copper refinery was established at Townsville by a MIM subsidiary, and Townsville became a key beneficiary of Mount Isa's productivity. The mine had already been sending parcels of ore to Australian smelters and refineries at Port Pirie and Port Kembla. Copper production resumed in 1953 with a new copper smelter. The US owners ratified a major expansion program in 1956 after further vast reserves of copper were found. Huge underground access shafts and a continually growing estimation of known ore made Mount Isa the largest underground mine in Australia, and one of the largest in the world.

In all areas of its operation MIM was supported by the Queensland government. This relationship was established from the beginning of the mine in 1924, and the passage of the *Duchess to Mount Isa Railway Act* the following year. Government support for mining was common in most states of Australia, but the Queensland government also offered financial assistance in the late 1920s, including large loan guarantees. The assistance included a company-friendly industrial relations regime, and very lax occupational health and safety oversight.[38] It was not that Queensland governments of various political complexions were not unconcerned about the wages and conditions of Mount Isa workers – it was just that they were subservient to the need to keep production rolling. Any break in operations was a significant threat to the entire state, as the mine produced 60 per cent of the state's mineral wealth by the 1960s.

The workforce grew from 3671 in 1961 to 5288 in 1971. Across all areas of the mine and smelter there were increases in production. From 1961 to 1970, blister copper production almost doubled, silver production almost trebled, and the amount of silver-lead-zinc ore mined grew to over 2 million tons per annum.[39] This drove population growth: from 7433 in 1954 to 13,358 by 1961. It almost doubled again, to 25,497, by 1971.

Arriving

The extent of population growth post-1945 meant that large numbers of people were arriving in town all the time. The moment of arrival looms large as a key experience and memory, and first impressions stayed with residents for many years. They also influenced the likelihood of staying; of shifting from the world of itinerant, mobile worker to the world of locally engaged resident.

These first impressions were sometimes unfavourable. A distant smoke plume rising from the copper or lead smelter stacks is usually the first sign of the approaching town. The newest lead smelter stack, completed in 1978, is 270 metres tall. Glimpses of it can be seen over 30km away, and its emanations can be spotted from even further away. A straggling industrial and commercial area, with engineering firms and truckstops, greets the traveller arriving from the west, along with the mining lease itself, which is full of industrial buildings, mullock heaps, and different coloured ores and spoilage. From the east, as the Matilda Highway runs from Longreach to Cloncurry through Winton, Kynuna and McKinlay, the flat plains give way to weathered, undulating country of red soil and low clumpy scrub.

Julius Kruttschnitt described his first impressions after a long journey by train from Sydney to Townsville and then on to Mount Isa:

> My first impression [in 1930] was nothing but dusty roads and nothing above the height of a one storey building. There was nothing there. The railway yards were quite barren and the settlement was back I suppose about half a mile from the railway.[40]

The expansion program carried out under Kruttschnitt changed the townscape considerably. The new smelter and furnace buildings were

Mount Isa

an inescapable presence. Margaret Jones, who came to Mount Isa from Brisbane in 1957 as a newly-wed, remembers her first impressions of them: 'you could see these big tall buildings, like a big ship'.[41] The streets and roads were still red dirt. The development of residential suburbs after 1950 also set up quite a contrast to the earlier brown calico tents and dusty corrugated iron buildings.

There were also cultural and climatic differences for English and European migrants to negotiate. Noel Addison, an English migrant, arrived in early January 1961 at Mount Isa Airport: 'Stinking hot! 5th January. The airport was just a tin shed and there was no-one coming for me.' Sigge Lampen described his Finnish sister's first visit to the town:

> She drove a car from Brisbane to Cairns and here, and camped. And they were like, they were like shocked because the distances were so long and there were no trees, and there were no houses and there was nothing. She explained it – she is a European at heart.

Lampen also indicated the difficulty of capturing the essence of the town:

> It is just a mining town where people drink and get carried away, and work, and you know ... Look, in this town there are so many nationalities. I think it would be hard to explain Mount Isa to anyone.[42]

The story of the difficult arrival is commonly expressed through the local understandings of distance. Jones recalled her English friend bringing her mother to Mount Isa:

> Barbara said to me one time: Mum's coming out. I am going to drive to Townsville to meet her. So she drove out and meets Mum. And Mum says, oh well we'll get in the car and drive to Mount Isa.

> How far away is it? Oh only about ten hours' drive. Well Mum didn't know what to think! She didn't have any idea.

These stories are often followed by an increasing appreciation of, and reconciliation with, distance and locality. As Jones recalls of her friend's English mother:

> Well once she got out here and found her way around she enjoyed the place. She didn't want to go back. This is what happened to a lot of people that have come from other countries.[43]

Modernist dreams

Behind the stories of arrival are less defined stories about hopes and dreams. Industrial modernity promised a new world, a better life. It affected all sections of society.[44] The scale and speed of Mount Isa's growth seemed to make these dreams sharper, clearer. European migrants were especially responsive, as their lives had been profoundly changed by war, and by the slow and difficult reconstruction period.

Betty Addison emigrated to Australia aged 20, with her husband Noel, and moved to Mount Isa in 1961. In response to the question, why did she choose Australia, she told the following story:

> I can remember in school, in Geography, seeing a picture in a geography book. It was full-page picture of a beautiful field of corn, bright blue sky and a man on a tractor, and underneath it it said "Beautiful Queensland".[45]

When she and her husband attended an interview with an Australian immigration official, she 'said to this man "Queensland". He said "Fine", so we ended up in Queensland [laughter].'[46]

Other stories have an almost whimsical quality, with the dreams of youth realised by adults. Kalle Nurmi had always dreamed of escaping the cold winters of Finland. Chain migration, family support and economic necessity were all part of his migration story, but behind it lay that childhood dream:

> From an early age I hated the winter – not the work but the winter time. From a very early age I had been saying I'd been born in the wrong country. Winter time for me was more like a nightmare.[47]

Working for MIM

It was paid work that brought these workers to Mount Isa. Progress was measured in good wages and a good lifestyle. Sandy Jones returned to Mount Isa in 1957 to work underground with advice from his father not to 'return as a tradesmen because tradesmen were only on a flat rate. Not like those underground.'[48] Noel Addison, who came to Australia as a 'ten-pound Pom' on the *Arcadia* in 1957, found work at Mount Isa in 1961, attracted by the high wages: 'When they [MIM] told me how much the wages were I said put me down straight away. The wages were three times more than they were offering in Brisbane.'[49]

With the expansion program of the 1950s, Mount Isa became one of the largest and most mechanised mines in the country. It had underground and open-cut mining for copper and silver/lead/zinc, and lead and copper smelting. In 1961 MIM started building a new shaft, known as K57, a huge circular concrete-lined shaft which was officially opened by the Queensland Premier, Frank Nicklin (later Sir Frank), in 1966.[50] The shaft could transport three times more men and equipment into the mine than the older Man and Supply shaft.[51] Its size allowed the move to diesel-powered vehicles, and they replaced railway-borne transporters and ore haulers in the early 1970s.[52] Trucks, utes and mechanical diggers could simply be driven to the work areas. Images of K57 became iconic representations of the underground operation.

Underground in the mine the temperature was usually over 30°C but it could be as high as 60°C if there was oxidising ore in the immediate area. Ventilation throughout the mine was controlled by a huge system of shafts and bulkheads. Some areas were dusty and contained acid fumes after explosive charges ('shots'); working in the smelter, the crushing plant, or in transport and storage had its own challenges.[53]

The rich copper deposit beneath the Black Star lode, proved in the 1930s and 1940s, was joined by another even richer body of copper-bearing ore located in the 1950s, beneath the Rio Grande mine.[54] Once again, governments played a crucial role in MIM's expansion plans, with a railway line to Townsville, funded by state and Commonwealth money, completed in 1959.

In 1951, at the height of the price spike in export commodities caused by the Korean War, the basic weekly wage in Mount Isa was £22/5/5, which included £13/10/- in bonus payments. The minimum wage for Queensland in early 1951 was around £7/12/-.[55] The wages for contract miners needed to be high to attract workers to the site. However, the high wage rates need to be balanced against the cost of living in Mount Isa. The isolation and the climate added to the cost of housing, fresh food, and heavy imported items such as fuel, furniture and electrical goods. In 1951 fuel was selling for 4/5¾ a gallon at Mount Isa; the equivalent Brisbane price was 3/3.[56] The *Woman's Day* sent a reporter to Mount Isa in 1965 and found widespread complaints from women about the cost of staples such as bread and milk, and larger purchases such as furniture. Even electricity supplied to company-owned houses cost twice as much as in many capital cities.[57]

Pat Mackie, leader of the miners during later industrial disputes, wrote that wages had dropped, on average, from £29 to £25 per week between 1952 and 1959. Mackie's own work history exemplifies this story. After an unsuccessful trial as a contract miner, he moved to a timber gang working underground. He was earning £8 a day on piece rates in the early 1960s, but as MIM changed work patterns and rosters, his wage dropped to £6/10/-.[58]

Furthermore, wages could not be predicted. A mining crew could be put to work in an area where the returns per ton of ore worked were low, and thus be paid low wages. Alternatively, they could be denied a contract altogether, particularly if they were identified as 'troublemakers' by the shift boss. Until 1960, when the Industrial Court set the bonus at a flat £8, the bonus varied according to the price of the commodity.

Despite these caveats, high wages were available to certain male workers at Mount Isa. The caveats merely balance the regular MIM-sourced commentary which exaggerated the wages boom. The same is true of the company's emphasis on heavy drinking and illegal gambling at Mount Isa in its evidence before the Queensland Industrial Court: their claims were true in terms of the three hotels of dubious reputation, but the company exaggerated the problem in order to undermine the case for wage increases.

But the labour market extended beyond MIM and its contractors. The town employed large numbers of workers in retail, administration and the public service. As Mount Isa grew it became an increasingly important regional centre for health care, and government services and programs.[59] The commercial sector would prove important during the industrial dispute of 1964–65, siding with the miners, and providing them with significant financial and moral support. The relationship between the mining workforce and local businesses was a close one.

The mining operations – and the smelter, especially – also shaped the visual and aural aspects of the town, and gave the air an odorous quality in the right winds. Sandy Jones came back to Mount Isa in the late 1940s to visit his parents, who ran a pie shop in the main street, next to the Commonwealth Bank: 'Fumes were very strong. When the fumes [came], there'd be a big blue cloud come out of the smelter.'[60] The other thing Jones noted, and this can still be felt, is the noise and vibration from underground blasting. Residents lived with a constant multisensory reminder of mine operations. On quiet nights the hum of the smelter is a palpable, almost subliminal, presence, punctuated by occasional sirens and reversing hooters. The other emanation from the smelter and mine was airborne lead. In the 1970s and 1980s, locals would come to recognise this as a crucial problem.

Suburban growth

Expansion was vital, to house the town's growing population. MIM championed a suburban expansion program, but not through any ideological commitment to home ownership; it was all about labour supply and retention. In 1953 it was estimated that the annual turnover of the workforce was 47.5 per cent. This was serious, because many jobs required training and, ideally, many years of experience.[61] The answer was suburban development and home ownership, through a new partnership between the company and the state. Helping workers achieve the suburban dream gave the company a distinct advantage: workers with some local financial attachment – that is, a mortgage – were less likely to move to another town. As we shall see, this local attachment also led to stronger, more militant union organisation.

By the early 1960s, Mount Isa was a modern, permanent, suburban community. However, there remained a need to close the social gap between Mineside and Townside that had opened up by 1939. The gap was a product of MIM's need to attract labour quickly and to provide an acceptable lifestyle for its better-off staff and skilled workers. After the war, a more subtle, long-term strategy was needed. Another very pragmatic reason drove the company's move to abolish Mineside – payable ore had been found in 1960 near this residential area. Mineside's days were numbered.

Soldiers Hill, so named because the area was the location of an army camp during the war, was MIM's first major experiment in suburban development, built in the early 1950s. The first of ten Co-operative Housing Societies was established. The first houses at Soldiers Hill were simple one-bedroom fibro homes, with the company guaranteeing low-interest loans. Later homes were larger, set on piers for improved ventilation. In 1953 MIM negotiated with the state government and the Cloncurry Shire to surrender 70 acres (28.3 ha) of its mining lease for housing development. Such land swaps were the basis of new suburban growth, including one of the largest postwar suburbs, Parkside. Parkside was situated between the Leichardt River and the

Duchess–Mount Isa railway line, and was the site of Mount Isa's first multi-storey residential project, Parkside flats, built in 1969.[62]

The houses were first made of tin sheeting, and later weatherboard. Margaret Jones arrived in Mount Isa in 1957. Of her first house in Deighton Street, she remembers:

> It wasn't a flash house. It was a dirt floor. Tin walls and the windows were shutters that I pushed out. ... Over the back of the hospital. No hot water. In summertime when it was so very hot the pipeline was above the ground so when it was very, very hot I had hot water in the pipeline [laughter].

Securing the basics to fit out a house was difficult, especially as local prices were high. As Margaret Jones recalls, 'We had to buy the furniture to put in it. We bought one piece at a time, as we could afford it.'[63]

Straight after the war, with the mine producing strong returns, money was invested in an appropriately impressive house for the managing director. Casa Grande, in Nettle Street, Mineside, was completed in 1949. It is an impressive Spanish mission-style house, designed by Brisbane architects but also possibly influenced by Kruttschnitt himself, given his time in California, Mexico and Arizona. The house, which is still intact and well preserved, was set on higher ground than the surrounding home sites and has a commanding view from its two storeys. It is built around an enclosed courtyard; there is a tennis court and extensive gardens, which are a strong symbol of resources and wealth in this arid zone. The house was the centre of staff social life and also entertained visitors from interstate and overseas, including company directors and Queen Elizabeth II in 1970. It was the Kruttschnitt home (with his second wife, Edna) until 1953, whereupon it was taken up by the new managing director, George Fisher.[64]

Only 5km from Casa Grande were the tent houses, constructed by MIM in a desperate attempt to provide accommodation for workers in the town's earliest days. Sixty tents had been erected in 1930 in a site known as the 'tent town' area. This area had an outbreak of diphtheria in July 1932, with one child dying and several others hospitalised.⁶⁵ The tents were converted into 'tent houses' in 1934, with MIM providing the materials and employees donating their labour. These simple timber structures had canvas roofs and iron sides. The floors were dirt or board.⁶⁶

Exclusions

Not everyone moved into the new Mount Isa suburbs. There were still a number of tent houses even in the late 1960s. The penultimate tent house was demolished in 1991, and the final one, in Fourth Avenue, has been preserved by the National Trust. Aboriginal people remained at a distance from the high-wage boom economy of the white miners and their families. The Aboriginal Reserves established in Cloncurry and Camooweal discouraged, but did not entirely stop, Aboriginal settlement in Mount Isa. A camp developed to the southwest of the town and became widely known as Yallambee. The population of this area was highly variable, but in 1976–77 there were about 59 Aboriginal people at Yallambee with another 94 living in a council-run caravan park.⁶⁷ Other Indigenous people were scattered throughout the town in what researchers Mark Moran and Dell Burgen called 'dispersed settlement'.⁶⁸ A 1976 report by the Mount Isa Welfare Council estimated the total Aboriginal population for Mount Isa at 3000.⁶⁹

There was pastoral work for Aboriginal men and women throughout the region, but the diverse and often labour-scarce economy of Mount Isa also offered employment opportunities. Aboriginal employment in Queensland after 1897 was regulated by the Chief Protector, and union-won award wages were not available to many. Wages, where they were paid, were placed in bank accounts controlled by the local protector, usually a policeman. However, many Aboriginal workers ignored or avoided these regulations. Historians Jim Hagan and Rob Castle estimated that some 80 per cent of Aboriginal pastoral workers were outside the regulated system in 1962.⁷⁰ Work was available, but not at the wages enjoyed by non-Aboriginals.

Mount Isa

Mount Isa had numerous attractions that were not always available to Aboriginal people, including the annual rodeo (from 1958), cafés and theatres, and late-opening hotels. The strong local rugby league competition provided a reason for talented Aboriginal players to move to Mount Isa, and perhaps to eventually move to one of the Brisbane clubs. The town's claim that it dominated other northern Queensland towns in rugby league, made in 1929, did not come to pass until 1969, when Mount Isa won the Foley Shield, the holy grail of the North Queensland Rugby League Association.

Local Indigenous history has now entered into the public history of the town, as Indigenous people now make up about 15 per cent of the town's total population. Mount Isa is a centre of regional Indigenous life. Indigenous families move into and out of regional centres like Mount Isa for employment and education, for access to health care, and for social and cultural reasons. In particular, movement between Mount Isa and Aboriginal settlements such as Dajarra and Domadgee, or towns such as Camooweal and Boulia, both of which have high proportions of Aboriginal people, is a common and significant feature of regional population mobility.[71]

Another distinctive component of the Mount Isa population was the Finnish community. Finns had a long tradition of hard rock mining and many, sometimes with their families, emigrated to northwest Queensland. The first Finnish workers arrived as early as 1924. By 1939 there were reportedly 200 in the town.[72] By the postwar period there was a vibrant Finnish community, with a Lutheran and Pentecostal church, a Finnish Association, and a Finnish-style café in the main street. The café was a piece of cosmopolitanism that many city-based visitors did not expect to find in Mount Isa.

Kalle Nurmi was a Finnish migrant who came to Mount Isa in 1960. He was born in Uleva, a farming community in central Finland, and he grew up on a farm. His migration to Mount Isa was influenced by his older brothers, who had come to Australia cutting sugarcane in 1959. His first impression of Mount Isa has a distinctive Finnish rendering: 'It was very much a new experience to me. It was something like what you feel when you walk into a sauna.' His brothers had moved to Mount Isa in 1959, so when he arrived he lived in a caravan in his brothers' front yard for a year then moved to rented accommodation. In January 1968 he started as an underground miner on twice the rate of pay he had previously earned working on the surface.[73]

The Finns were joined by migrants from Italy, Greece, Yugoslavia (as it was then known) and other nations in Central and Southern Europe. Out of the total population of 12,516 in 1966, some 35 per cent were born outside Australia. These were some of the highest rates recorded in our sample towns: a testament to the extent of Mount Isa's growth. The largest groups were from the United Kingdom and Ireland (as one would expect), at 1594; but there were also 446 from Germany, 229 from Italy, 225 from the Netherlands, 145 from Yugoslavia, 140 from New Zealand, and 114 from Asia. All of these groups had the usual preponderance of males over females.[74]

The extensive economic and town development in the 1950s bought political changes in the 1960s. Gone were the progress associations of the earlier decades – the town was less about appeals to outsiders and more about taking control of its own destiny. This meant the gradual eclipse of Cloncurry. Mount Isa was moved from Cloncurry Shire to the Shire of the Barkly Tableland in 1962 and the Shire was renamed the Shire of Mount Isa. The base for the Royal Flying Doctor Service was moved from Cloncurry to Mount Isa in 1964. In 1968 Mount Isa was declared a city. State members for Mount Isa – the seat was established in 1972 and included a large rural and pastoral area as well as the town – all became vigorous supporters of their town, drawing their power from a solid sense of identity and a common feeling of being left out of the mainstream of Queensland politics, with its focus on the coastal and southeast regions. As befitting an increasingly important and confident town, a local newspaper, the *Mount Isa Mail*, was established in 1953. It eventually absorbed the *Cloncurry Advocate*, another indication of the changing relationships between the two towns. But the spectre of MIM was never far away. The first editor of a local newspaper established in 1965, the

Mount Isa News, was a former public relations consultant for MIM. It is not known whether this unduly influenced the editor's approach, or whether the company had any clear role in the establishment of the newspaper, but the idea of a newspaper for the town was discussed with the MIM general manager.[75]

In the 1960s, a substantial housing commission program began: in 1966, these houses (in the new suburb of Sunset) cost between $8400 and $9400. This exemplified the company/state relationship that was distinctive to Mount Isa. MIM had entered into agreements with the Commonwealth Housing Commission 'under which a percentage of the Commission homes were reserved for rental to company nominees'.[76]

Suburban growth brought new facilities. The town gained a new 85-bed hospital. The road to Alice Springs and Darwin via the Barkly Tableland was sealed by 1967; the Flinders Highway to Townsville was fully sealed by 1976. By 1966 there were 700 company-built homes in Mount Isa. Of these, as one company publication noted, 'almost 200 are being sold to occupants through the home purchase society'.[77] The crowning achievement of outback suburbanisation was the Lake Moondarra scheme. This involved a huge dam, built by MIM, 16km from Mount Isa. It ensured a reliable water supply for the mine and township from 1958, and was the beginning of a recreational area for boating and swimming (it included an 'inland beach'). From its completion in 1962, the beach and recreational area were popular – it explains the high boat ownership rates in the city. Those unaware of Lake Moondarra found it hard to understand where these boats might be used.

By the 1960s Mount Isa was a town that was here to stay. A 1967 company publication spoke directly to concerns from would-be newcomers about the transient nature of mining towns. According to the company, the town 'has outlived its temporary era and moved into a period in which the emphasis is on permanence – in solid buildings, good roads and modern facilities'. The white race had proved itself worthy. The anxiety which came with early development and an uncertain future, and which generated so much activity, had given way to confident assertions of permanence.

The 1964–65 dispute

But permanence had consequences that MIM did not foresee. A strong financial and emotional stake in the locality provided the bedrock for worker militancy in the 1960s. This was given increased significance by the sheer number of waged workers in Mount Isa. The 1956 Census records that 3090 males and 437 females were 'employees', out of a total workforce of 3752: some 94 per cent of the workforce were wage earners.[78] Mount Isa was, numerically speaking, a working man's town even before any mobilisation around that identity.

It is not my intention to cover the 1961 and the 1964–65 disputes in great detail: they have been well covered by other scholars, and the focus on one or two high-profile industrial incidents is part of an approach to town history of which this book is critical. Nonetheless, looking at the disputes from other perspectives does throw light on the nature of local society. The miners' leader, Pat Mackie, has written that in the lead-up to the 1964–65 dispute 'labour-management relations were going from bad to worse ... It almost seemed the Company was deliberately provocative, needling the men at every opportunity.'[79] First, local management were under pressure to justify their expensive development program, and second, the new bonus systems introduced in 1959 were linked more clearly to production targets.[80] These management changes produced a new period of industrial disputation.

In 1961 there were amendments to Queensland's *Industrial Conciliation and Arbitration Act* which meant that the Industrial Commission could no longer hear an application for an increase in bonus payments (though it could still reduce or curtail them). This caused great concern among the men, and the ensuing dispute closed MIM for two months. The AWU did not officially support action, but a Combined Unions Council at Mount Isa did. The men returned to work with no real solution to their concerns. The strike was ended by an order-in-council issued by the Queensland government which declared 'a state of emergency'.[81]

In 1964, in an effort to bring MIM to the bargaining table over the bonus and other concerns, miners *en masse* signed letters indicating that they would move from contract to wages employment, which was an option under their current award. There was a significant escalation in the dispute when AWU officials and MIM colluded to have Pat Mackie fired; he was subsequently ejected from the union by AWU officials. Mackie then emerged as a rank and file leader of the miners. He was a man with a long history of union activism, who was influenced by the principles of the Industrial Workers of the World (IWW). The company closed the copper smelter in an effort to prove that the move to wages work was part of a 'go slow'. The miners argued that there were plenty of raw materials available to continue production for many months.[82]

The Queensland government announced a state of emergency under the *State Transport Act* on 10 December 1964. Under the Act it was unlawful to question the operation of the Act. The state of emergency order required a return to work the following Monday under threat of a large fine and six months' gaol. From 13 December 1964 all union meetings were completely open to the public and the press. AWU officials were booed off the stage and the men demanded that Mackie chair the meetings.

On 14 December 1964, state Cabinet conferred on MIM the power to sack workers who refused to work on the contract system. This was followed by an order-in-council in January 1965 which gave the police powers to search without a warrant, arrest without charge, and exclude certain persons from town if they believed those persons might threaten industrial peace. All underground workers were sacked and the mine was closed on 9 February 1965. The response to this and previous government interventions was a full-scale mobilisation of the community. A mass meeting was held on 15 December 1964, and 4000 attended, including the wives of miners and local businessmen. Hundreds of police were despatched to Mount Isa – a response that locals found particularly galling because there had been no violence at union meetings or on picket lines.[83]

A significant feature of the dispute was the division between the AWU executive and its membership. Lawrence Miller travelled from Sydney to Mount Isa in 1954, aged 29, after advice from his brother-in-law that there was work available. He started at MIM as a rigger, earning £8 to £10 per day. His godfather returned to work during the dispute and was branded a scab. Miller never spoke to him again. Miller felt that 'the problem with the AWU was they were a shearing union in a mining environment'.[84] Mackie, like Miller, saw the AWU as essentially a pastoral and agricultural workers' union:

> Most workers had long felt they would be better off in a miners' union such as the Miners Federation or the Workers Industrial Union of Australia.[85]

Granted, Mackie had a particularly critical view of the AWU hierarchy, but his views were shared by many workers, including Miller, and he did have significant support during and after the dispute.

The social history of Mount Isa from 1945 points to another dimension of the 1964–65 dispute. Mackie noted that it was married men with many years of service who were eager to prosecute the case against the company and against AWU officials:

> Keenest were many men with long years of service, with families and homes in Mount Isa, skilled miners, hard workers on contract who could not simply pull up stakes and leave in search of other work, and who had long denounced the AWU leaders' selling them down the river with the inadequate bonus agreements in 1959 and 1961.[86]

So home ownership both provided a strong financial commitment to place and made it difficult for workers to move to other fields. Suburbanisation and home ownership, seen as a bulwark against communism and union militancy in the Cold War era, actually encouraged the men to fight harder in this dispute.

The dispute built successful cross-class and cross-union coalitions in the town; it also brought in the broader labour movements of Queensland, New South Wales and Victoria. The miners of Broken Hill, through the Barrier Industrial Council, were especially supportive, sending strike pay levied from their members and delegates to offer moral support and advice. Local coalitions brought together social groups that had diverse interests or were not always in close co-operation. Local professionals and businessmen including a newsagent, a doctor, a café proprietor, two taxi drivers, and the owner of three electrical goods shops, funded Mackie's wage after he was dismissed.[87] At key moments the local Chamber of Commerce supported the miners and their union allies.

This contrasts quite strongly with Premier Nicklin's view. He suggested that Mackie was a troublemaker who managed to:

> upset one very strong, very moderate section of workers at Mount Isa, the Finnish community, and divide them among themselves and create a tremendous trouble in the area. And when things developed there Mount Isa became a Mecca for every communist in Australia.[88]

Nicklin's perspective is a powerful reminder of the broader political and cultural climate of the 1960s, which was shaped by the Cold War, the fear of communism, and the perceived or imputed threat to an 'Australian way of life'. Rather than community coalition, Nicklin saw discord and division.

Finally, a cross-class mobilisation of the kind that occurred at Mount Isa (despite Nicklin's view) was not possible without a strong sense of class identity. The large number of miners attending meetings and the high levels of support for the miners from other town-based unions, as expressed by the Mount Isa Trades and Labour Council, reveal the increase in union support and identification in the late 1950s and early 1960s. By this time, class was a significant element of local life. This class identity even transcended gender differences, for a time. The Housewives Association handled payments to workers and women attended the union meetings and picket lines.[89] It also helped overcome ethnic divisions in this most multicultural workforce. Once the rank and file, led by Mackie, took over local organising, miners of Italian, German, Finnish, Dutch and Yugoslav origin, among others, were brought into the life of the union through the provision of interpreters and a new policy of welcoming 'New Australians'.

After an eight-month battle, the men accepted a return to work in early April 1965. As Sandy Jones, a veteran of the dispute, recalled:

> We didn't get much out of it. Only a bit of respect for ourselves. We used to work three shifts here. Eight hours each. Now they work 12-hour shifts and all this crap.[90]

One new working condition that came indirectly out of this dispute was the offer of an annual return airfare to Brisbane for MIM employees and family members. The company made much of this in publications from the late 1960s and it became an accepted feature of local employment conditions.[91] In another example of unintended consequences, the free airfare became the focus of another major industrial dispute in 1995, when MIM sought to remove it.[92]

Stereotypes and realities

While people may expect the town to be populated by hardworking, hard-drinking miners and their long-suffering wives, this is a caricature that is rarely matched by local realities. Mining towns, including Mount Isa, have a cultural diversity that confounds visitors with its breadth and depth. The Finnish community is still present. Sigge Lampen, a Finn who grew up in Sweden, arrived in Mount Isa in the early 1980s, having lived in East Africa and Canada before settling in Australia. Lampen does not seem a typical mining town employee:

he met his future wife in the Philippines and married her in Mount Isa in 1985. A miner and talented builder, Lampen shuns the local hotels, but revels in the cosmopolitan mix of his own household and the freedom, space, physical heat and human warmth that the town gives him.

Margaret Jones sees change in the community in the last 20 years. She feels that more recent arrivals are less inclined to join organisations, and that the voluntary sector has experienced a significant shortfall. New residents are not joiners, or their work patterns make participation much more difficult:

> Everybody seemed to be happy to join in and be involved. Nowadays it is not as easy as that ... New ones that come in ... seem to be more involved with their children or going to work.[93]

MIM moved to 12-hour shifts in 1991. A number of respondents identified this as a significant change to the work–community relationship. Many sporting and service clubs suffered as afternoon training or evening meetings were made impossible by the new roster system.

There is also an increasing subtlety in the public presentation of the town's history, best seen in the 'Outback at Isa' exhibition, a presentation that uses visual, aural and tactile elements to present a sophisticated, lively, and at times critical, local history. The 'Outback at Isa' exhibition is one of the most effective historical presentations in any museum or interpretive centre in Australia, another fact that upsets the common notion that the regions are backwaters, always trailing capital city cultures. Mount Isa's modern history shows a sophistication which coexists uneasily with older industrial and pioneering histories; it is not easily categorised or caricatured.

While accepting these complexities, Mount Isa's history remains dominated by the story of a harsh arid landscape transformed into something 'normal'. As powerful as this image is, and indeed it is clearly marked in the landscape, there is no escaping the fact that The Isa is in the middle of a desert.

Mining Towns

In the end it is the juxtaposition of suburb and desert, of the town and the vastness of the desert sky, that gives Mount Isa its distinctiveness, at least to outsiders. For locals it is what it is.

7

KAMBALDA
Modernity, environment and experience

On the dusty-red, open woodland country of the goldfields of Western Australia is a town with two histories. Kambalda's first history, based on gold, began in the late 1890s. It ended within a decade. Prosperity had come and gone, as is common in the mining industry. The town was pronounced dead, only to come back to life. Kambalda's second history, based this time on nickel, began in the 1960s. This second history is the subject of this chapter. The 1960s Kambalda was a product of that decade's town planning and labour management strategies. It was a quintessential 'company town', and epitomised the powerful modernist claim that industry could overcome the environment and create an urban oasis in the outback.

Kambalda is a valuable contrast to the late 19th century mining towns where there was little planning. The whims of the market and wild rumours on the field drove these towns' chaotic and rapid development, with governments straining to keep pace. Even Port Pirie and Mount Isa, partly company towns, did not match the approach taken at Kambalda. Living in the modernist dream, however, turned out to be much like living in any other mining town. The dissonance between the design and the experience, between the plans and the reality, between the city of ideas and the town in the outback, reveals many of the paradoxes of modernism itself.

Twentieth-century mining and industrial towns were exemplars of

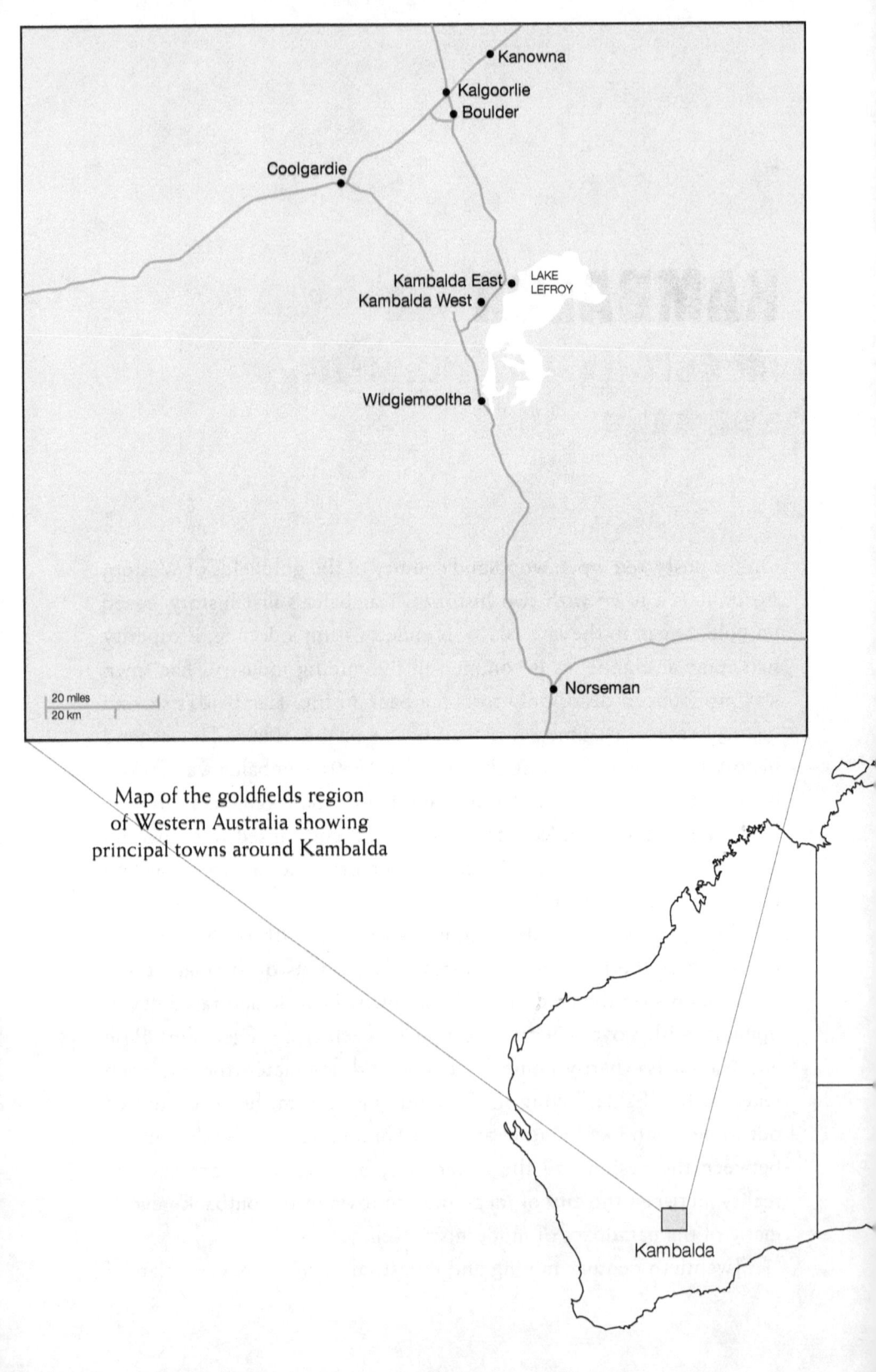

Map of the goldfields region of Western Australia showing principal towns around Kambalda

Kambalda

a mode of thinking that privileged science and 'progress'. In the workplace, employers remade working conditions in order to create capitalism with a human face; the same approach was also applied to other areas of working-class life that were seen to be in need of reform. This thinking was at the heart of attempts to create company towns. BHAS at Port Pirie and MIM at Mount Isa grappled with these same issues, and also created, for a time at least, company towns. BHAS grafted a company sensibility onto an existing port and rural service community. MIM had created a town divided in two at Mount Isa, with Mineside and Townside signalling that the project was only ever half-finished. Kambalda, sitting above massive nickel and gold reserves, was different. It was an unprecedented opportunity to design an ideal town from nothing. The original goldmining town had been abandoned and new plans treated the site as a greenfield one, where the skills and knowledge of town planners, architects and social scientists could be brought to bear. The approach used many features of earlier schemes but it also had new elements, elements specific to the 1960s and, significantly, to the location.

In Kambalda, the summer sun bakes the air and earth. Hot winds roll in from across the plains and willy-willies weave red strands across the horizon. As in Mount Isa, spring, winter and autumn weather is moderate, with the days bright and warm, the nights cool to cold. Perth is some 600km to the west. The shimmering waters of Esperance are 350km to the south. The nearest major regional towns are Coolgardie, 55km northwest, and Kalgoorlie, the most notorious of them all, the upstart town that surpassed Coolgardie with the discovery of the 'Golden Mile' in 1894. Sixty kilometres to the north, Kalgoorlie sits like a fixed compass point, holding everywhere else in place.

Kalgoorlie provides the best long-run climate data. There are almost nine days per year when the temperature goes above 40°C, while the mean maximum summer temperature is 33.6°C. The mean minimum is 5°C in July. The average annual rainfall is 268mm (some sources give a Kambalda figure of 242mm), most of it falling in late summer in abrupt and heavy downpours; mid-winter rain is light and persistent.[1]

The original Kambalda goldmining operations were on the edge of Lake Lefroy, an impressive salt lake to the south and east of what would become the future town site, running over 50km south and up to 16km wide. The lake was named after Henry Maxwell Lefroy, whose exploration in 1863 piqued the interest of those in search of pastoral land. The first Europeans to travel in the region lamented the lack of rivers, good grazing country and permanent water sources. Yet like so many places in outback Australia, after rain the country comes alive with new growth and wildflowers. This temporary fecundity could deceive the visitor. Expeditions led by Lefroy in 1863 and Charles Hunt in 1864 were both conducted after heavy rains. North of Coolgardie Lefroy reported, somewhat optimistically, 'that few finer sheep runs are to be found in Australia'.[2]

In 1866, Hunt named the lake after Lefroy but found little reason for optimism. At nearby Parker Hill he surveyed the scene: 'from there obtained, embracing all points, an extensive view of the most wretched country'.[3] This was the European perception of this country for the next 25 years. The region is home to the Galangu people, who often, along with many other goldfields Aboriginal people, acted as guides for European explorers.

The first Kambalda

The area first gained significance for Europeans in December 1896. Wretched country became gold country. Percy Larkin was an experienced prospector associated with the gold finds at White Feather (later Kanowna), 23km northeast of Kalgoorlie. He discovered traces of alluvial gold in the gullies and dry creek beds that ran from the top of Red Hill south towards Lake Lefroy. As one observer wrote, 'the precious metal here is found among the angular stone and red earth'; this stone and earth is still readily apparent in the area today.[4]

Larkin's find provoked a rush (a small part of the gold mania that swept Western Australia in the 1880s and 1890s), attracting 400 to 500 men, mostly from Coolgardie. After a rich start, the easily won gold was soon worked out and larger operations with deeper shafts replaced the smaller claims. The Red Hill (WA) Gold Syndicate acquired a number of leases and began mining alluvial gold set in underground river sediment in 1897, at a site close to the edge of Lake Lefroy. The town was officially gazetted on 10 December 1897 and laid out by the government surveyor, W. Rowley, who selected the name 'Kambalda', apparently because 'it sounded like a pleasant name'. However, there is a similarity between the name and words of local Aboriginal languages, many of which were the source of similar names, such as Kalgoorlie and Kanowna.

There were 400 to 500 inhabitants and seven or eight stores not far from the mining leases, near the present-day Kambalda East. One contemporary visitor noted that the town site was on 'a sheltered flat at the foot of the longest gully'.[5] Another praised the beauty of the town, noting 'the usual canvas shanties' and 'the most substantial buildings in the place being the two hotels – built of corrugated iron'.[6] As in other goldfield towns, a Progress Committee was formed and became the de facto local government, representing the interests of businessmen and storekeepers in particular. Ore grades began to decline in 1904. By 1905, the company, a London-based operation, was struggling. It ceased mining altogether in 1906.[7] The town that had sprung to life so quickly was more or less abandoned by 1907. Many other towns in the region also grew overnight, only to decline, and sometimes disappear altogether, in a few years.

Coolgardie was the base from which most of the gold discoveries of the 1880s and 1890s were made, but it was soon overtaken by Kalgoorlie. After 1907 Kalgoorlie strengthened its regional dominance, eventually joining with neighbouring Boulder to create the twin cities of Kalgoorlie–Boulder in 1989. By 1911, the combined population of Kalgoorlie and Boulder was 19,605. It was much the same until after World War II, then increased to almost 30,000 after 1986.[8] Kambalda's relationship to Coolgardie still remains strong, because it was incorporated within the Coolgardie Shire. It is now one of the two towns in this small shire.

From the 1930s to the 1960s the area continued to show tantalising signs of mineral wealth. The most famous example was the discovery in 1931 of the Golden Eagle nugget at Widgiemooltha, 40km south of Red Hill. Farmer-turned-prospector George Cowcill had been attracted to the region in the late 1930s, and initially worked a site not far from the old Kambalda township.[9] Cowcill discovered metal-bearing ore. After teaming up with John Morgan, the two returned to the area in the 1950s, managing to attract the attention of Western Mining Corporation (WMC).[10] Founded in 1933, WMC was a Collins House company built on significant West Australian-based gold leases. In the 1950s and 1960s, WMC was mining nickel, bauxite and iron ore as well, so Cowcill and Morgan's 1964 ore samples from Kambalda, which contained reasonable levels of nickel and copper, were appealing. The four samples they submitted assayed at 0.7 per cent nickel and 0.5 per cent copper.[11]

Most accounts of Kambalda's 'second coming' focus on Cowcill and Morgan and the WMC staff who supported their hunches. This individualist account leaves out the broader political economy of the mining industry. Cowcill and Morgan, as venerated pioneers and prospectors, are matched by Patrick Hannan in Kalgoorlie, Arthur Bayley and William Ford in Coolgardie, Charles Rasp at Broken Hill and John Campbell Miles at Mount Isa. Yet nickel had been identified in the area as early as 1897. What always kick-started major development of a find was appropriate market conditions, a supply of capital and labour, and a supportive state government.

The second Kambalda

Postwar Western Australia placed the highest possible priority on industrial and mineral development. The political economy ensured that the state operated as a major facilitator of, if not partner in, capitalist development.[12] It was much like Queensland in this regard. New projects were encouraged and made viable by significant state investment in infrastructure, especially transport. In the goldfields that infra-

Kambalda

structure extended to the provision of a water supply, crucial to both human health and the mine machinery.[13] The famous water pipeline from Mundaring Weir in the western foothills near Perth to Kalgoorlie was completed in 1903. The railway from Perth to Kalgoorlie arrived even earlier, in 1897. Transport links east to Port Augusta (South Australia) were complete in 1917; those south through Norseman to Esperance began operating in 1927.

The price of nickel and the size of the orebody brought the decision to start large-scale mining and build a new town at Kambalda. In 1950, the nickel price was US$992 per ton. By 1960 it was US$1630, and it rose to US$1940 when production commenced at Kambalda in 1967.[14] The find at Kambalda was highly significant in Australian and world terms. While initial exploration suggested an orebody of around 700,000 tons returning an average nickel content of 5.2 per cent, later exploration defined the available tonnage as being around 1.4 million tonnes.[15]

Australia's principal nickel mines and smelters are in Western Australia, and Kambalda was the first major nickel discovery, and operation, in the country. Nickel helped shaped the future prosperity of WMC. Just as Broken Hill profits flowed out to other mine and smelting locations a century earlier, nickel mining and smelting reaches out to Kalgoorlie, Kwinana and Esperance. The mineral itself is a valuable addition to stainless steel: adding nickel increases the alloy's resistance to corrosion. Nickel can add strength and conductivity too. As a component of high-end alloys, nickel is widely used in the aviation industry, and in chemical processing. Most nickel is found in complex sulphide ores, the same type that gave the Broken Hill companies such grief at the beginning of the 20th century. By the 1960s, however, smelting processes had advanced considerably, and in 1972 WMC built a nickel smelter just outside Kalgoorlie. The smelter used Finnish flash smelting technology from Outokumpo Oy. The process was by this stage a highly efficient and well-tested method of separating nickel from its mined ore. The Kalgoorlie smelter produced nickel matte, which could be further refined by WMC's Kwinana refinery or exported through Esperance to Canadian refineries.[16]

In 1966, Kambalda's second life and second history began. From 1964 to 1966, WMC carried out drilling in the Red Hill area. Journalist Peter Laud reflected on the notion of a new life for the area when he wrote: 'the place which was until a few months ago a dead loss has woken up'.[17] By the middle of 1966, about 20 men from experienced drilling crews in Kalgoorlie, were employed on the Kambalda drilling rigs, living in caravans and tents. Historian Norma King interviewed the principal driller on those early rigs, Jack Lunnon:

> We had a bad run beforehand and we didn't expect too much to come of it. Quite honestly we didn't expect optimistic results at Kambalda. Our ultimate find was a pleasant surprise.[18]

The first major mining area was named the Lunnon Shoot. WMC continued naming the key oreshoots after the leading drillers involved in their discovery.

By August 1966, prefabricated living quarters for single men had arrived at the town site. Ambitious plans for a substantial, sophisticated company-built town were under discussion, but they were modified and delayed by two important factors. First, production had to start up as soon as possible for financial reasons, and second, there was uncertainty over the exact location of the orebody.[19] There was always a possibility that major town infrastructure would be built over areas that later proved to have mining potential – this fate befell the state electricity commission town of Yallourn in Victoria's Latrobe Valley.[20] So the early preference was for temporary housing. These units were 'timber framed, asbestos clad, fibrous plaster lined dwellings with timber floors and corrugated asbestos roofing'.[21] Asbestos was the building product of the age.[22] Inside, every unit had four bedrooms, each with an inner-spring mattress or a divan bed, a wardrobe, a desk and an armchair. Graeme Hunt, Kalgoorlie correspondent for the *West Australian*, reported that 'Men living in the village will have comfort unknown to those who have previously lived and worked in isolated mining sites in Western Australia.'[23] The new town had arrived, for the most part on the back of large trucks from Perth. The new Kambalda was self-consciously presented as a break with the past: previous goldfields towns had failed and disappeared back into the wilderness, but this town would bring a new modernist aesthetic and presence to the red earth near Lake Lefroy.

Kambalda's revival was also a contrast to the relative stasis, even decline, of Kalgoorlie. As outlined in the Mount Morgan chapter, gold prices did not increase markedly after 1945, and the industry required Commonwealth assistance. The Great Boulder Company which worked the Golden Mile returned only modest profits, slowly shedding jobs. South African author Elspeth Huxley visited Kalgoorlie in 1965, and found 'a lack of confidence in the future' revealed through the 'lack of new buildings and the shabbiness of the old'. In this context, WMC's plans for Kambalda were a powerful symbol. Huxley wrote about this too: 'But hopes of revival have been kindled by a promising new find of nickel at Kambalda, only 30 miles away…'[24] In 1972, Norma King, who grew up in Kalgoorlie, wrote:

> The twin towns of Kalgoorlie and Boulder in 1965 B.N. (Before Nickel) were rather depressing places to live in. There were voices of gloom everywhere … The general uneasiness and lack of confidence were reflected in the appearance of the towns. They looked like rather shabby old ladies who had known better days, as indeed they had.[25]

Nickel would be the saviour of the struggling goldfields, at least in the minds of those who were aware of the prospects of another ore being found. As King wrote (and the use of 'we' here marks this as a local perspective):

> When it was realised that we had other treasures, besides gold, hidden in the ground, the goldfields took on a new life, and we felt as if we had experienced a last minute reprieve.[26]

The importance of the Kambalda find to Kalgoorlie was realised in a more tangible way when WMC and the state government signed an agreement allowing the nickel smelter to be built in either Kambalda or Kalgoorlie. As local resident and former WMC employee Will Manos noted:

> This was an olive branch, so to speak, from Western Mining and the Kambalda community to Kalgoorlie, who were struggling at the time. In reality they could have built the nickel smelter right next door to the nickel mill. They felt that by putting it closer to Kalgoorlie that would provide some financial gain for Kalgoorlie as well as employment opportunities for some of the blokes in Kalgoorlie whose jobs weren't worth much in regard to the price of gold at the day.[27]

Sure enough, Western Australia's 1968 *Nickel Refinery Act* provides that 'the State will make available and sell and grant to the Corporation and the Corporation will purchase from the State for a smelter such land at Kambalda or Kalgoorlie'.[28]

Another context shadowed the arrival of a new planned town. The goldfields region is dotted with ghost towns, the remains of failed mining ventures. Nowhere else in Australia is the finite nature of mining town development so dramatically evident; many of these old towns are now part of tourist and heritage trails, though in many cases there is little, if anything, left to see.[29] The surrounding environment was seen as relentless, malevolent. The bush appeared to be devouring the towns that had once been the wellspring of great hopes and aspirations. Kambalda, with its modernity and conveniences, heralded a fight-back against nature.[30]

By November 1966, WMC had planned 50 houses for its employees, with a town site located within the boundaries of the original Kambalda, though the streets and lots were to be redrawn. By August 1967 there were 300 residents – some 50 families. WMC leased the entire site from the state government and was in full control of its residential, commercial and industrial development.[31] The WA Premier, David Brand, officially opened the operation in September of that year. The main shaft, constructed on the Lunnon Shoot site, was named the 'Silver Lake Shaft', following a competition publicised by the company magazine. The winner was Mary Norman, who, with her husband Arthur, was reportedly among the first five families to settle in Kambalda East.[32] The company policy of involving staff and residents in naming parts of the mining operation helped create a local idiom. It was a shrewd way to help staff and residents 'own' the new environment.

The development of the mine site precipitated major industrial and infrastructure projects. In 1970, the Kalgoorlie to Esperance railway was diverted to run through Kambalda, and work on the nickel smelter on the outskirts of Boulder began. By 1979, there were eight mines operating in the region.

WMC's engagement was comprehensive, covering planning, construction of roads and houses, and the provision of the water supply, sewerage, garbage collection and telecommunications. This arrangement was formalised in an Act of parliament which set out the rights and responsibilities for WMC and the WA state government. The Act was one of many contractual arrangements between the state government and private companies known in Western Australia as state agreement Acts. These have governed the private sector/government relationship since the 1950s, particularly in the mining and petroleum industries. Under the *Nickel Refinery (Western Mining Corporation Limited) Agreement Act 1968* (WA), WMC agreed to construct the nickel refinery by 1972, and the state government agreed to offer land for sale or lease for the refinery and for the town site, among other things. More specifically, for Kambalda the WA government agreed to grant to WMC land for 'residential, professional, business, commercial and industrial purposes and the provision of communal facilities at Kambalda [via] a special lease or special leases'. The lease or leases would be for 21 years at a modest rent, with WMC having an option to buy in all cases.[33]

In Kambalda, WMC had decided to control the labour process and the social sphere. This was not the preferred WMC strategy; it was the result of factors beyond their control. Labour would not come unless living conditions and local amenities were of a high standard. As one unnamed senior manager ('KN') noted in October 1969, in terms of town development:

> the municipality could not, even if it would, move at a speed satisfactory to us. We are then placed in a position where we must take matters into our own hands ...[34]

The town building project was a necessity, 'irrevocable capital expenditure' as KN called it, but current taxation law did allow tax deductions for about one third of the expenditure.[35] Not only was the shire unable to deliver for WMC. Resident Manager J.B. Oliver felt Western Australia's Town Planning Department's plans showed an 'inadequate regard for the local topography and soil conditions'. After the initial Kambalda East plan prepared by the department in October 1966, WMC took over all town planning, led by architect Geoffrey Borrack and landscape consultant Jean Verschuer.[36] Kambalda East had 300 blocks housing 1200 workers and their families, together with 100 in the 28 two-storey townhouses, and 150 in the caravan park. The old grid plan for the 1890s Kambalda was reworked, though care was taken to preserve a historic well from Charles Hunt's expedition and a pepper tree planted by the original settlers. The contractors' employees' camp, housing some 900 men, was 'removed from the town areas for social, industrial, traffic and aesthetic reasons'. By the end of 1967 a supermarket, a recreation centre, and a post office were opened at Kambalda East.[37]

Kambalda West

The town has two histories, two lives, and two parts. As had been feared, WMC uncovered more reserves of nickel-bearing ore close to Kambalda East, so future town development was planned about 4km to the west of the township. Kambalda West was developed between 1968 and 1973. Its curved streets and preservation of on-site vegetation signalled a new approach to town design.[38] The extent of the orebody was now known with more certainty and a town of 10,000 was planned: Kambalda East, Kambalda West and a subdivision north of Kambalda West (which was never developed).

The Kambalda West subdivision revealed the influence of the garden city and town planning movement. Typical of the Radburn design principles which influenced so many postwar suburbs, large access roads ringed the area, and residential and park areas were clearly separated from industrial and civic areas. The central commercial area and a light industrial area were placed at a distance from the residential lots.[39] Natural slopes and creek beds were incorporated into the design – this also characterised the uranium mining town of Mary Kathleen in Queensland. Another Collins House venture, Mary Kathleen was designed in the mid-1950s and was widely regarded as a successful and progressive company-sponsored development.[40] Most blocks in Kambalda West faced north. (Houses on west-oriented blocks were the subject of some complaints and were not highly sought after.) As well as the preserving of local vegetation, the company also established a planting program to beautify the new town and help suppress the dust that could sweep so unforgivingly across it.

The company houses in Kambalda West were generally of a better quality than those in Kambalda East. They were three- or four-bedroom brick and timber fibro-clad houses – with tile roofs for the senior staff. The majority of the houses for wage workers had fibro-asbestos roofs. The company chose cement floors after experiencing failing wooden floors in Kambalda East. The houses were partially open plan in style, with a combined lounge and dining area leading to a small galley kitchen. Many early residents found the lack of storage in the kitchen and laundry difficult. The interior was a mix of cement brick and fibro, with heating provided by an open wood fire in the lounge. The efforts to preserve trees extended to adjusting house alignments so that trees could remain. This accounts for the erratic house placement – which, ironically, is at odds with the carefully designed road pattern. It also gave 'The West' a greener, softer edge than 'The East'. Many oral history respondents noted that cutting down a tree in a WMC house was a 'sackable' offence. The company supplemented the tree preservation policy with the provision of a generous water supply; some residents developed extensive gardens, and even lawns. In the more water-conscious 1990s, WMC offered incentives for residents to remove their lawns.[41]

Kambalda fits all the definitions of a 'closed' company town. As well as being the dominant employer, WMC controlled all aspects of local residential development, from collection of garbage to the design and maintenance of the houses. In the 1970s, for example, a maintenance gang performed household repairs on company houses as well as general repairs on roads, stormwater, and signage. Electricity for the town and the mine was supplied by a company power station. At the heart of Kambalda was 'the company'. Its control over the town mirrored its control over the production process: from exploratory drilling to mining and maintenance, and from processing to transportation and smelting. WMC dominated in the way the Mount Morgan Company had at Mount Morgan, and the Mount Lyell Company had at Queenstown.

As the major employer, working conditions and experiences at WMC shaped the local labour market. By the end of 1969 WMC employed 1335 workers. This peaked at 1704 in August 1977, just before the first of the major retrenchments.[42] Miner David Stokes remembered conditions underground as 'terrific', certainly much better than on the surface:

> I discovered right from the time I went underground [that] they treated men like men. Generally speaking there was a very different type of man that was down there, especially as far as the bosses were concerned.[43]

As at Mount Isa, it was warm to hot underground. In the drives and stopes the temperatures were 30°C to 34°C depending on the depth and the nature of the ventilation. Industrial relations, according to Stokes, were 'quite good':

> The industry was booming, so every year there used to be a new round of negotiations for new price increases, especially for those on contract work like the machine miners, truck drivers, bogger operators and so on. For the ones on fixed rates there was a good increase ... around 8 to 10 per cent.[44]

As we have seen, a constant feature of mining and industrial town labour markets was the skewing of paid work opportunities towards what were seen to be male occupations. Underground and surface employment at WMC – at the deepest levels of the mine, in the primary crusher or in the town maintenance gang – were all men's jobs. However, WMC understood that many of their male employees were married and that there had to be some paid work for women. Some women moved into small business and retail, and others did find work at WMC.

The three 8-hour shifts shaped the rhythms of the community. Retailers knew the fortnightly pay days for day labour, and also the monthly pay days for staff workers. In good years, extra money came to staff at Christmas through the nickel bonus, and large purchases were often scheduled for this time of year.

National and international attention

By the early 1970s Kambalda and the new wonder metal, nickel, had a distinctive profile in the region and nationally. *The Women's Weekly* published a number of stories on the new town, commenting that Kambalda was now a competitor to Kalgoorlie:

> But Kambalda is new, with all mod cons, such as air-conditioning, smart new houses, up-to-date working conditions with the latest

machinery for working the mines. It appeals to young married couples.⁴⁵

The commodity itself was subjected to widespread financial speculation, leading up to the infamous Poseidon float in 1969–70, when shares in a nickel prospect at Windarra, northeast of Kalgoorlie, peaked at over $280 per share. This coincided with a high in nickel prices – almost US$7000 a ton. A precipitous decline began in February 1970, with many investors losing substantial amounts of money.⁴⁶ As one market economist drily noted, 'Given the relatively low ore grade, the extraction costs were very high and, thus, the mine was no more than a break even proposition.'⁴⁷

Kambalda, particularly Kambalda West, was further highlighted in 1973, when UNESCO sponsored a major conference on 'New Towns in Isolated Settings'. A member of the Australian delegation to the 1970 UNESCO Conference, Professor O.A. Oeser, suggested a seminar on 'man and his social environment'. UNESCO agreed and a suitable site in Australia was chosen. That site was Kambalda: one-third of the participants were to be accommodated in the town itself and the remainder were to stay in Kalgoorlie. Kambalda was chosen because it would enable participants:

> to compare the haphazard, unplanned town [i.e. Kalgoorlie] with one in which not only every building, facility and amenity had been built to a plan but in which the concept of environmental preservation had been a part of the plan from the very beginning.⁴⁸

The pairing with Kalgoorlie is a further demonstration of the way town identities are constructed within active relationships. Kambalda was both the economic saviour of the region and the orderly, modern, younger sibling of the chaotic ageing goldfields town. The seminar participants came from government, industry and academia. The final report from the seminar included studies of a number of planned towns in Australia, including Kambalda. The conference generated a body of valuable evidence covering the first six years of Kambalda's second existence, and was an example of the kind of sustained analysis that brought new knowledge to bear on planned community life.

The key players were social scientists, from academia and industry. Oeser himself was a highly regarded political and organisational psychologist. He became foundation Professor of Psychology at the University of Melbourne in 1970, and produced an important body of work on isolated towns and the mining industry.[49] The two men carrying out the Kambalda studies were academics who were in unusual positions. P.J.R. Redding was a lecturer in sociology from Preston College of Advanced Education, but was also listed as a 'Research Associate' of the Human Relations Unit (HRU) at WMC. Similarly, J.K. Austin was a lecturer in administrative studies at Caulfield College of Advanced Education but also a 'research associate' of the HRU.[50] The HRU was a specialist unit set up in 1970 to study company town function and effectiveness. WMC had clearly identified academic expertise as crucial to the effective resolution of their social and labour problems. The HRU was the vehicle they used to host, and shape, that research.

The final Kambalda study was jointly authored by two senior managers at WMC. L.C. Brodie-Hall was an executive director of the company, and J.B. Oliver was a former resident manager at Kambalda (1967 to 1971) and now general manager of Kalgoorlie Lake View Pty Ltd. Brodie-Hall and Oliver wrote that the HRU was to 'enquire into the structure and process of management, the relations between management and the community and the structure and process of community living'. The terms 'structure' and 'process' are a striking illustration of how the language of sociology and psychology had become a part of the language of management.

None of this is to suggest that these studies were not legitimate research. Qualitative data informed the composition of wider survey questions, populations were sampled, and response rates were measured. The results, including findings that reflected poorly on the company, were reported with apparent openness. What was significant was the combination of academic knowledge and the experiences of a discrete population. BHAS had pioneered the development of in-house expertise to address labour management issues, and WMC had pursued partnering with academic researchers even further.

In his research for the final Kambalda report, Redding surveyed 37 married couples living in Kambalda; some were staff, some were waged employees. One third were born overseas, mostly in the United Kingdom. The Australian-born group were more likely to be from regional areas in Western Australia. The new inhabitants, particularly those with plans to stay more than a few months, were overwhelmingly from the west. There were few, if any, Aboriginal people in this group. In the 1970s WMC did not encourage Aboriginal employment. Many would not have got jobs at the company, and therefore would not have been eligible for company houses. While there were 4689 Aboriginal people resident on the goldfields in 1996, there were few in Kambalda.[51] There is also a possibility that Lake Lefroy is seen as an undesirable place by Wongatha people because of a story of a Wongatha girl who disappeared on the lake.

A picture of a group moving from one employer to the next through many different towns emerges strongly from the WMC research. The principal reason cited for moving to Kambalda was work-related for men: in particular, security of employment for waged men and career opportunities for staff men. For women, the overwhelming reason for the move was to follow their husbands; most then performed a home duties role. The median length of residence for these new arrivals was only 18 months. Two-thirds of the men had worked previously for at least four employers, sometimes in different states. Only 22 per cent of the sample wanted to settle in Kambalda, and 45 per cent could specify the period of time they would stay. This latter point accords with Will Manos' experience: his family arrived in 1975 with a 'five-year plan'. As Redding noted, the 'striking things about these findings are the low proportions who would settle, the degree of uncertainty about the future, and the fact that intentions do not vary over different groups'.[52] Kambalda was viewed as a place to live for a period of intense work and higher wages; this is still a dominant mode of thinking in mining towns. Few moved there for lifestyle reasons.

One thing that the survey did not pick up was the role of family and friends in the move to Kambalda. A number of oral history respondents mention this. The Manos family was encouraged to move from Darwin to Kambalda (via Perth) by grandparents. David Stokes, too, recalls that a relative who worked for WMC helped him secure his first job in the town.[53]

Living in a planned community

Memories of arrival are again clear. Norma Worthington recalls her first day in Kambalda:

> I arrived on the last Sunday in September in 1971 and, on the following Friday, I turned 39 and there were school sports and I drove the car in ... And I drove into the [oval] ... and the cars were all around the oval like a footy match. They don't do that anymore cause of something. The numbers have increased so much. And I saw the entrance to the oval; when I looked around at all the people and all the cars that were there ... I didn't know one person here ... but that altered.[54]

Initial adjustment could take some time, but it was easier for families with school-age children – the schools were a venue for meeting and mixing with other locals. That was particularly effective for mothers who were left at home during the week.

David Stokes arrived in Kambalda in October 1970 as a married man of thirty-five looking for a new start. His parents had moved around Western Australia, especially in the northwest, when he was a child. After doing his National Service in the army, and living in Queensland, he and his wife came to Kambalda. Stokes' sister and her husband had told them about it. Stokes recalled:

> First impressions were very good. A lot of the houses were still being constructed at that time. They were no real shops over on this side [Kambalda West] ... The only shop over on this side was a converted house that they used until the plaza was built.

Securing work wasn't a problem. Stokes worked in the WMC store and later went underground as a miner.[55]

The company's housing, much praised in newspaper reports and often mentioned in company annual reports, was less well regarded by Kambalda residents. While 85 per cent of the sample preferred the detached homes to the duplex apartments on paper, this changed when couples had to live in them. The summer heat and the winter cold were major issues, as were the waiting lists – 3–6 months. The houses were draughty, not dust or insect proof, and unlike the artists' impressions drawn up in Melbourne, the backyards were usually bare. Despite the efforts to preserve trees, shade was still at a premium in the early phase of town development.[56]

Kambalda East houses were not as well constructed or as appealing as the newer homes in Kambalda West. Some of the Kambalda East homes were 'transportables' set on piers; their wooden floors warped in the heat. Some had also been roughly treated as a continuous round of short-term contractors moved through them. For staff families used to more salubrious homes, these rougher houses were a shock.

There were other issues too. Norma Worthington noted that:

> You had to wait three months for a house in those days. My mother thought this was like Stalag 29 or something when she came here. It was all new. There were no gardens ... They're all company houses. And there was, you know, no greenery; this was a new street. This was a new street and it was very drab.[57]

Still, both wage and staff families appreciated WMC's investment in building this new town. For wage employees the rent was $5 a week, which included gas, water and electricity. Staff families paid $6 per week, which included gas, water, electricity and firewood. The common designs of the homes gave a design motif to the entire residential area, especially in Kambalda West. Whereas the houses in other towns covered in this book showed variations across styles and periods, Kambalda's one theme still shapes its urban landscape. Kambalda, in true modernist fashion, arrived ready-made, with form and function shaped by materials and topography. The houses remained in company hands until after 1986, when was there a conscious policy to sell the homes to employees. This was the first sign of WMC gradually withdrawing from full control of Kambalda.

Putting down roots

New arrivals steadily added to the town's population. By late 1971 the population was around 4500, not including the few hundred itinerant, short-stay workers in the caravan park and single men's quarters in Kambalda East. In 1974, for example, it was estimated that 160 families were living in short-term accommodation – flats, duplex apartments and the caravan park – in Kambalda East.[58] Tourists or senior managers were accommodated at the Kambalda Motel, which had opened in Kambalda West in April 1970. The primary school in Kambalda East opened in 1968 and the Kambalda West Primary School opened in 1972. By 1974 the shopping mall, in the central hub of Kambalda West, was complete, as was the local high school. By 1977 the population had increased to 5600.[59]

This was the heyday of Kambalda. As David Stokes recalls: 'The shopping centre was crowded on a pay Friday. There were shops all the way through ... and there was a really happy atmosphere.' The motel often had a band on a Saturday night and WMC and other town organisations ran meetings there constantly. As Stokes recalled, it was the hub of the town's social life:

> Every Friday night the function room at the motel was set up as a

smorgasbord. People would go down there with families. There'd be music, you know, a local band. It was really great.

For him, these years were very good:

> The houses were a better quality than those we were used to, coming from Queensland ... Overall, I was actually proud to say that I worked for Western Mining in those first few years ... They were treating people right. They spent a lot of money building this town. Nickel was doing very well for the first few years.[60]

For Norma Worthington too, after the initial period of adjustment, the early 1970s were wonderful years. Her children were enjoying school. She took a job at the WMC Laboratory. Her husband enjoyed his work, and occasionally played the piano at the Kambalda Club, in Kambalda East.

As families matured and children grew up, WMC remained a consistent and highly regarded employer of young men. The apprenticeships available covered all industrial and commercial trades. Worthington's children all secured apprenticeships with WMC. Will Manos started with WMC as a 15-year-old apprentice in 1986:

> It was a completely closed mining town and they were self-sufficient in their maintenance of the plant and of the community itself ... As we left school, friends of a similar age to myself were employed in trades in many different fields. It was a wonderful thing that you could pretty much choose whatever type of trade that you wished to get into if your grades were good enough coming out of school, and Western Mining could accommodate you in that.[61]

WMC returned strong profits in the early 1970s, peaking at over $21 million in 1971, allowing it to invest in the community and in its own future workforce.[62]

Kambalda

This investment included funding for local sporting clubs and facilities. The Kambalda Football Club was inaugurated in 1969, and moved to the new Kambalda West Oval in 1971. As if to reflect community vibrancy and strength, the Eagles won the Goldfields Grand Final in 1976. To many in Kambalda, defeating Kalgoorlie- and Boulder-based clubs that had been in existence since the late 1890s meant the town had come of age. The decline of the Eagles in the 1990s and 2000s, equally, was seen as evidence of a community in decline: the new shift arrangements made regular commitments such as training very difficult.

The common housing stock meant that social and class differences were not as obvious as they were in other company towns. WMC was adamant that there was to be no 'nob's row', or identifiable management area, in Kambalda. Nevertheless, differences were still present. Senior staff did have four-bedroom houses, double garages, bigger blocks, and larger pantries and living areas. The company employed gardeners who worked on senior staff homes as well as around the town. Staff houses had electric water heaters; wages houses had wood-fired heaters. Staff also had subsidised petrol. The staff also had their own sporting competitions, in which wage employees were not allowed to participate. In certain areas, staff Christmas parties excluded the waged employees who worked in the same section.[63]

Despite the challenges and the occasional mistakes, it is clear that the very highest levels of WMC management were engaged in the process of community building. They earnestly discussed the pros and cons of particular house designs, sought advice from academic experts, and invested significant amounts of company money. One early estimate suggested that building Kambalda would cost WMC $5.5 million – a level of investment that is at odds with present-day practice in the mining industry where a drive-in-drive-out or fly-in-fly-out workforce not only removes the need for town-based investment, but actively undermines regional communities.

Economic downturn

The optimism generated in the new town in the 1960s, and the focus on designing a new, workable community in the early 1970s, were replaced by familiar concerns about job losses and community decline from 1977. In this sense, Kambalda was little different from the older towns. Ambitious plans were predicated upon a strong economic base. Even the addition of gold production, from 1980, failed to protect the company from declining commodity prices and international recession.[64] In September 1977, WMC indicated to the Minister for Mines that 154 staff and 516 waged employees were to be dismissed.[65] The recession of the early 1980s had an impact too. The minister at that time, P.V. Jones, claimed that the Australian nickel industry was not as badly affected as some – the Canadian industry in particular – and indeed sales of refined nickel and nickel matte continued, albeit at lower prices. Across the WMC nickel operations, which included the mine and smelting facilities at Kalgoorlie and Kwinana, the reduction in the workforce would be through 'natural wastage and by the recently announced twelve-month wage freeze on staff salaries'.[66] It was a change in the industrial landscape, and would profoundly affect the community of Kambalda. The wage freeze made salaries less competitive with other mine sites, and so led to the less permanent members of the workforce leaving town.

In 1991 the financial press reported that WMC had put an 'ultimatum' to its Kambalda employees. The company wanted to see more flexibility in work practices, especially in working hours. 'In the event that the company's position is rejected,' managing director Hugh Morgan warned, 'workers currently surplus to requirements will be offered employment in their classification at Revenge mine on continuous shift work.'[67] The AWU resisted the company's move towards seven-day-a-week continuous mining and the men often 'finished up on the oval' – the local term for stop-work meetings, always held at the Kambalda East Oval. The case was lost by the AWU in the Industrial Relations Commission. In response, WMC's share price increased and the miners at Kambalda went on indefinite strike. WMC, the miners argued, had failed to provide adequate health and safety assurances in relation to the proposed new work roster. WMC put their redevelopment plans for the nickel mines on hold, waiting for the AWU to agree to their plans. This was no idle threat. In late 1991 WMC announced that the redevelopment plan was halted and 150 men would be retrenched, thus effectively outmanoeuvring the union. This was followed by increasing moves by WMC, from 1996 on, to contract out mining operations to smaller companies.

The roles of Sir Arvi Parbo (managing director from 1974 to 1986 and board member until 1999) and Hugh Morgan (managing director from 1986 to 2002) cannot be underestimated. In 1991 the left-leaning press characterised the Kambalda developments as 'New Right on the Bash'.[68] Parbo and Morgan were quite different from the earlier generation of Collins House managers. The postwar period was dominated by managers such as Gordon Lindesay Clark (later Sir Gordon), who was WMC chairman from 1952 to 1974. Clark was thoroughly Collins House, a member of the major clubs in Melbourne, associated with the University of Melbourne as a benefactor, adviser, and one-time lecturer, a resident of Kooyong, and on the board of the more progressive Broken Hill companies such as Broken Hill South. One of his final acts of significance for WMC, other than writing the company history, was to create the WMC-Lindesay Clark Trust Fund 'to assist mining communities' in 1979.

Thereafter we see a slow turning of WMC policy and approach to mining communities, most clearly from the early 1990s. Parbo and Morgan were less interested in community development and philanthropy, and both were influenced, from the mid-1980s, by the 'New Right' and the growing tide of business activism through groups such as the H.R. Nicholls Society, of which Morgan was a founding member. They sought to reform and overturn work practices which they saw as limiting the business. The move to contract mining, and the more aggressive style in industrial negotiations, signalled a decisive end to WMC's company town experiment. By 1996 the company that had created a town in an arid zone desert, and planned for aesthetic and environmental values, had disappeared entirely.

Just as it disengaged from and disowned the communities it created, WMC itself eventually lost out to the heartless market – victim of a corporate takeover from none other than BHP in 2005. The ghosts of the men who had pioneered the Collins House group, from W.L. Baillieu through to Colin Fraser and Lindesay Clark, would have turned in their graves.

There were more recessions. In 2000, the Minister for Energy reported that 500 jobs had been lost in the five mines at Kambalda in the previous two years.[69] In the 1996 Census, the combined population of Kambalda East and Kambalda West was 3598.[70] By 2001, Kambalda West had recorded one of the largest population declines since the last Census: by 20.8 per cent.[71]

WMC sold the Kambalda mine sites to BHP in 2001. One of the first acts of the new owners was to change the names of the original ore bodies. In a conscious strategy, BHP obliterated the contribution of WMC and its workers to the development of the area. They placed their own logo over major signage, even memorials. One high-profile example was the re-badging of the huge nickel pots WMC had placed on the Kambalda–Boulder road just outside Boulder. For decades these pots had represented the modernist industrial dream being pursued by WMC and the Kambalda townspeople. Now they signified an acquisitive capitalism that had no ties to community and history. The pots were no longer a memorial to mining and industrial development; they were just a public relations opportunity.

WMC's policy of involving locals in the naming process had been a successful one. It had created a series of locations that were invested with cultural and social meaning, and that were well known and widely appreciated locally. BHP's renaming and asserting ownership over a past that it had purchased, not experienced, created much anger towards the company. Residents overwhelmingly regarded the renaming as a cynical exercise.

WMC's retreat from community building heralded a new approach

Kambalda

to mining company/community relationships. From the 1990s WMC had subcontracted mining operations to smaller companies, and the Coolgardie Shire eventually took over the administration of the township. The newer companies are smaller, and less committed materially and emotionally to one particular place. They tend to use a fly-in-fly-out or drive-in-drive-out workforce. This more diverse ownership structure has had to navigate through the rise and fall of global markets after WMC's withdrawal. From 2009 there has been a turnaround again, and recent reports suggest a strong revival of the mining operations that now ring the town. As I note in the Postscript, the extent to which this mining revival will benefit the town is an open question, as the new dominating companies have a different approach and different ways of operating.

The promise of modernity was most strongly expressed at Kambalda between 1966 and 1977, but it foundered in the difficult years that followed. As promising as the subdivision plans and the artists' impressions of homes were, they never quite matched the reality. All progress and development came with consequences, unmet expectations, and what German sociologist Max Weber called 'disenchantment'. Many aspects of Kambalda's history and culture are familiar to those who have lived in Kalgoorlie, Mount Isa or Broken Hill. Metal prices dictated the size and shape of the local labour market. Retrenchments and job losses played out in ways familiar to those in other mining towns. Pollution, environmental control and rehabilitation in Kambalda were not greatly different from what occurred in the older mining towns.[72] WMC itself, the company that had built and pioneered the town, not only withdrew, but was taken over by BHP.

Yet life in Kambalda is different from life in other mining towns. The planning, the housing designs and the greening project have all shaped its urban environment. Modernity reanimated the bush, returning an older abandoned town to life and adding the accoutrements, and conveniences, of the late 20th century world. It brought a creative power too, transforming the landscape and establishing an urban community and industrial infrastructure – arguably with its own beauty.

Mining Towns

Beyond the modernist dream, residents developed their own connections and meanings. Kambalda is now fondly regarded by those who have stayed. They are as engaged in their town as residents of any other small town. The sense of belonging shown by residents turns out to have been more pugnacious and long-lasting than the once all-powerful company that built the town, which is now gone altogether.

Postscript

The imaginary map of the Australian continent looks different now. The capital cities are still there and play a crucial role, but the six towns we've discovered in this book loom large. These towns no longer lie beyond our historical imagination, their inhabitants poorly represented in the 'national' story of Australian history.

The experiences of these townspeople remind us that other parts of the continent are shaded with rich knowledge and history too, rather than being empty spaces on a map. Formerly blank areas have distinctive environments and populations, making place and identity – together with movement and migration – central concerns of the 20th-century Australian experience.

In the late 19th century mining towns such as Broken Hill, Mount Morgan and Queenstown, the company-dominated Mount Isa of the 1920s, the coastal community of Port Pirie, and the company-designed Kambalda of the 1960s, local societies took root. Organisational and civic structures were established in a remarkably short space of time – in isolated locations and in the face of harsh conditions.

We see this lively culture through regional newspapers, where they exist, as well as churches, friendly societies, progress associations and trade unions. We see new towns and their inhabitants going about the business of creating a social world that gave form and meaning to their everyday lives. In the interviews conducted for this book, respondents talked of the meaning and identity in their lives. The place where they lived framed their social lives and memories; it was a source of stories, and a site of identification. The weight of these recollections shows that social and civic activity was fired by an individual's sense of local attachment and affiliation.

Mining and industrial towns exemplify the two potentially

contradictory themes of locality and migration, and offer new insights into the role of the nation-state. People felt they belonged to a local place or region, rather than a nation-state, yet still identified as Australians. But we can't disregard the nation-state: the role of governments in facilitating mining and industrial development is clear. From the 1880s, colonial, and later state, governments spent vast amounts of money on transport infrastructure, for example. Mount Isa emerged in its distinctive form because of interaction between the dominant company and a pro-development state government. Queenstown maintained production towards the end of the life of the mine through generous state and federal government support. Governments also set up the arbitration regime and the regulatory environment, and enacted the legislation that governed municipal authorities.

The fixing of culture in these mining and industrial towns is all the more remarkable given their rapid periods of growth and decline. And people's local ties proved extremely durable, despite the forces that disconnected, severed or compromised them, and propelled them – willingly or otherwise – towards new places, new lives. The movement of vast numbers of both overseas and Australian-born migrants was one of the defining features of these societies during their boom periods. Yet locally focused energy, effort and identity acted as counterbalancing forces. The dissonance between movement and location, between migration and local attachment, is another common element of life in these societies.

The 'spatial fix', as geographer David Harvey expressed it, refers to the providing of necessary infrastructure (plant, equipment, housing, towns) for particular types of economic regime.[1] The companies BHAS, MIM and WMC in particular all sought to create towns in order to attract and keep labour. This meant investment in local infrastructure such as water supply, housing and transport. Where the state would not – or could not – provide, companies had to step in.

Companies did not play a lone hand in this process. New residents also had an interest in building a liveable society. The 'spatial fix' wasn't

Postscript

only raw capitalist investment; it also led to living, breathing communities that people wanted to live in. This shared investment in building local infrastructure took on a social and cultural life of its own – my respondents believed that local life was valuable and important.

Of course there were different experiences across the country. Life in central Queensland is not like life in Tasmania, in ways that go beyond climate. In these towns social and economic divisions are obvious, as are ethnic and gender divides. Geographical isolation did not obscure or reduce these divisions. Staff houses were usually not far from rough workers' cottages and hastily erected tents. A palatial general manager's residence was a clear sign of the local power structure. And while the staff enjoyed clubs with billiard tables and reading rooms, the workers relaxed in hotels and built their own worlds in trade unions, friendly societies, churches and sporting clubs.

What united these groups was the fact that they lived in the same town and shared the challenges of local life. Though their solutions were mostly sectional or group-specific, at times they did come together in broader coalitions, such as progress associations and representative deputations. Occasionally it was politic to ignore local differences and present a united front. The outside world, particularly city-based company boardrooms and distant state governments, had significant power over local life. Locals knew full well that outside forces could determine their fate. The failed or declining towns that surround all of the case study towns were shadowy threats of failure that drove local boosterism.[2]

But not everyone had the chance to share in this local attachment process. Itinerant workers (and their families – some brought their families with them) often experienced the downside of rapid change. They were the people most affected by the whirlwind of grand plans or disastrous stoppages. Certain places, such as short-lived mining towns on the WA goldfields, or situations, such as being on the fringes of the labour market and beset by poor conditions or unfortunate timing, brought back mainly negative memories.

The resources provided by local organisations or kin networks enabled some itinerant families to stay put, but once a crisis point was reached a family was cut loose. The clearest example of this was the Great Depression, which was simply a more extreme example of the regular movement of large numbers of workers and their families from one industrial and mining town to another. Every move meant having to start again, building new local ties and establishing new local networks. There was a strong practical reason for local engagement: it was how job opportunities, informal support, and other resources — like friendship — were shared out. For those living in temporary camps, or boarding houses, the urge to move to the security of regular work and the comfort of a permanent home must have been strong, but not everyone was able to make this transition.

For Indigenous people the arrival of miners and industrialists was not an opportunity; it was a threat to their way of life. Many towns actively excluded Indigenous groups, or allowed only marginal and often exploitative relationships. This picture changed in the face of Aboriginal activism and changes to company policies in the 1970s, but Indigenous groups still have very different life outcomes in these towns. A clearer picture of their experiences requires further research. It is apparent, however, that Indigenous people who were subject to state control and intervention did not have the kind of control over their attachment to land and kin available to many non-Indigenous people. However, though they were often forced to move, Indigenous people's cultural and emotional attachments have often survived.

In towns where the orebody was substantial or where downstream industrial processing was successfully established, a sizeable majority of the non-Indigenous community was able to put down roots. However, these preconditions certainly did not guarantee a vigorous local society. Instead, our evidence shows a lot of hard work, false starts, high hopes and wrong turns, as organisations formed and lapsed. Building local societies took effort and perseverance, and organisations had to survive the town's ups and downs, plus national and international crises and internal ructions and rivalries. These towns were most successful in the post-1945 period. Steady economic growth generated well-paid

Postscript

jobs for most male inhabitants. Workers enjoyed more permanence and continuity – at least until the 1970s, when the next round of economic downturn and community decline began.

We can ask big questions of small places, drawing them into intellectual and cultural debates about the shape and direction of modern society. In the lives of mining and industrial town residents, we see the forces of modernity in operation, but we also see people as resilient agents and bearers of social change. There are at least two sides to the story: modernity isn't simply an imposed force that sweeps a nation and its people before it. It is instead a fragmented process that occurs in pockets.

The people in these towns were part of societies across the country that sought permanence. Rapid change and instability, including new modes of working, challenged them.

The ways of life found in these towns were the last phase in a type of modernity that has passed. No longer will capital and the state enable a place, a town, a company project, to develop *in situ*. There were few corporate towns built after Kambalda in 1966, and the era of the large private town such as Broken Hill or Mount Morgan has long faded.

In the early 21st century, there is a more pragmatic, utilitarian approach to servicing an economy. We appear to be even more subservient to the needs of industry. While people still seek permanence and certainty, the political order and the ruling economic regime privilege flexibility, elasticity and mobility. All the towns studied in this book were closely tied to the needs and rhythms of mining and industrial production, yet members of these communities wanted an active life outside work, in the civic and social milieu. It was the creative and occasionally conflict-ridden tension between making a living and making a life that defined these towns, and gave them their distinctive cultural heritage.

As the mining boom proceeds, temporary camps and fly-in-fly-out workers become part of the landscape in isolated locations; we need to understand the implications of this. Imagine Australia without Mount

Isa, or Kambalda. It would be geographically more homogenous and culturally less diverse.

While historians are best at discerning past trends, perhaps we can project a future where there is less urban difference, where our homes are geographically separate from work – sometimes by thousands of kilometres – and where our social and civic sphere is shrinking. One respondent from Kambalda told a story of a man who had returned to town after time away and wondered why the town seemed so quiet. He was told, 'Everybody is either working or asleep.' That is one result of the era of 12-hour shifts.

The sheer energy and diversity of local society gave these places their vibrancy, distinctiveness, heritage. The towns stand even where mining or processing has shut down. As we have seen, history and heritage may well be one way in which these towns can build new and sustainable forms of economic activity that help keep them alive. Broken Hill leads the way in this regard with a well-established city-wide heritage strategy.[3] One legacy from the past may be more difficult to overcome: contamination from past industrial and mining practices, in the absence of comprehensive clean-up or risk-minimisation programs, can still affect the health of locals.

As newer mining companies reduce their commitment to the idea of building towns, workers are likely to have a very different relationship with the place they work in. Fly-in-fly-out workers will not build a Broken Hill or a Queenstown, or even a company town such as Kambalda. The temporary camps and dongas will probably not be the subject of nostalgic reminiscence, and are unlikely to leave any substantial material remains. Space has been truly mastered by capital, but at what human cost?

I believe that this book is just a small sample of what needs to be done. This kind of research requires large numbers of researchers, as its scope is well beyond the capacity of one historian. We need to encourage research students to take on projects in far-flung localities.

It may be possible to make structural changes to the system of graduate funding but there also needs to be a change in the research

Postscript

culture that privileges city-based projects. This culture implies that historical experiences beyond the city limits are somehow less significant and that smaller towns took their cues from what happened in the city. We need to change a research culture that celebrates a conference in London or Washington while scoffing at a field trip to Port Pirie or Broken Hill. This is not a question of choosing one over another, as I hope my own contribution demonstrates. This is also no call for a parochial history that is obsessed with the local and the national. We can use town-based historical research as a point of departure, exploring the local while also charting connections and intersections with the global.

If we are, as a nation, to come to terms with living in this remarkable continent, with all its environmental and psychological challenges, we need to embrace all its places. To the 'city' and the 'bush', we need to add the 'mining town' and the 'industrial town', and no doubt many other locations where history coalesces with, and is expressed through, place.

Those who live in these mining and industrial towns have no trouble describing their worlds. Elizabeth Firth, Will Manos, Margaret Jones, Sigge Lampen and the others we have heard from describe strong and active civil societies. These places remain sharp in their memories. By connecting with regional stories and vernacular traditions, by listening to those seldom heard, and by considering their place in a national story, we can begin to create a more encompassing and inclusive history of Australian places.

APPENDIX
The oral history component

The research that underpins this book included an oral history project. In an effort to understand a local or vernacular view of the past I interviewed a small number of residents in each town. Some were formal interviews which involved a digital recorder, consent forms and follow-up transcripts and release forms, while others were informal encounters in local libraries, museums and community centres. All of the 32 formal interviews were undertaken with the approval of, and monitoring from, the Research Ethics Committees at the University of Newcastle (2006 to 2007), and Monash University (2008 to 2011).[1] I also accessed existing oral history from the National Library's extensive collection. The interviews were not meant to be statistically representative; they were about tapping into a remarkably shared sense of local culture. There were approximately equal numbers of women and men. There were no respondents under 40 years of age, so whether a strong placed-based sensibility is a generational issue will need further investigation.

I employed a 'life history' approach, asking questions about when and where people were born and then roughly following their lives chronologically through to the present day. Most interviews lasted between 30 minutes and an hour and in some cases were followed by a short session the next day. I asked open-ended questions such as 'Can you describe your family home?', 'What were your memories of school?' and 'What was your first job?' Where possible I tried to let the respondents set the themes and tone of the interview. If a respondent offered a memory or observation which spoke to the themes of local identity or community, I asked clarifying questions. I was careful not to direct or dictate the interview. I asked every respondent to describe

Appendix

what kind of history they would write if they were writing a history of their town. Where respondents gave their permission, the recordings and transcripts have been deposited in the University of Newcastle Archives and can be accessed by researchers.

While I secured appropriate approvals from respondents to use their names in any published work, I have used pseudonyms in order to protect their privacy. In a few cases, I also altered some details of their public positions or roles, so that they are not so easy to identify. These are small and often close-knit communities where it is not always easy to place one's views on record, so I am very grateful for the willingness of respondents to contribute. I trust I have done their memories justice, though I acknowledge that this book is ultimately my interpretation of their town's history. Given the commitment and enthusiasm that surrounded their engagement, I am sure this book will not be the last word, and I look forward to the interpretation presented here being extended, clarified or questioned.

Criticisms of oral history are often heard at history conferences, but are rarely applied with such vigour to written sources, which all have similar issues of accuracy and reliability.[2] On the whole I found a strong correlation between various forms of oral and written evidence, where I could triangulate sources of different types. So, for example, respondents at Mount Morgan recalled how the town changed in the 1970s, and the Census data, as well as ructions at the mine in that decade, provide both a context and a partial explanation for such views. The remarkable Syd Wall of Mount Morgan recalled many of the men who had died in the open cut and his recollections were backed up by newspaper searches. Respondents from Mount Isa remembered a highly multicultural community, and some were exemplars of migration stories and new migrant experiences. All of this makes sense in terms of the demographic data and the extent of the postwar boom in that town.

Mount Morgan and Broken Hill were valuable case studies because local researchers have conducted extensive oral history programs there in recent decades. I was able to compare that evidence with the material I gathered, and I found consistent themes across a range of projects, methods and chronologies. This gave me greater confidence in the oral evidence, as did comparison with oral history from the National Library. The National Library material was also particularly useful for tracking the views of those who had moved on from these towns. It provided important dissenting views.

Much more can be, and has been, written about memory and history, but if we start from the position that this history attempts to see the world as the respondents see it, then the question of establishing an objective account is less relevant, though not unimportant. Besides, all respondents were earnest and genuine in their efforts to describe their lives and their communities. I saw my task as to listen and respect their memories. Ultimately, the nature and shape of those memories, their tangible place-based qualities, was an inspiration for the historical analysis that I have presented.

Notes

Introduction

1 Interview with Elizabeth Firth. Further details about these interviews can be found in Appendix 1.
2 Interview with Will Manos.
3 See especially Catherine Nash, 'Setting Roots in Motion', in D. Trigger and G. Griffiths, *Disputed Territories: Land, culture and identity in settler societies*, Hong Kong University Press, Hong Kong, 2003, pp. 29–52.
4 J.W. McCarty, 'Australian Regional History', *Australian Historical Studies*, vol. 18, no. 70, 1978, pp. 88–105; Alan Mayne (ed.), *Beyond the Black Stump: Histories of outback Australia*, Wakefield Press, Adelaide, 2008, pp. 6–10; and Alan Mayne and Stephen Atkinson (eds), *Outside Country: Histories of inland Australia*, Wakefield Press, Adelaide, 2011, pp. 1–10.

1 The global and national context

1 A. Ford and T.G. Blenkinsop, 'Combining fractal analysis of mineral deposit clustering with weights of evidence to evaluate patterns of mineralization: Application to copper deposits of the Mount Isa Inlier, NW Queensland, Australia', *Ore Geology Review*, vol. 33, 2008, p. 435.
2 B. Kingston, *The Oxford History of Australia, Vol. 3, 1860–1942: Glad, Confident Morning*, Oxford University Press (OUP), Melbourne, 1988, pp. 38ff.
3 *Australians: Historical Statistics*, Fairfax, Syme & Weldon, Sydney, 1987, p. 133.
4 D. Meredith and B. Dyster, *Australia in the Global Economy: Continuity and change*, Cambridge University Press (CUP), New York and Melbourne, 1999, pp. 59–63.
5 Peter Cochrane, *Industrialisation and Dependence: Australia's road to economic development*, University of Queensland Press (UQP), Brisbane, 1980, p. 10.
6 Phillip Lipton, 'A History of Company Law in Colonial Australia: Legal evolution and economic development', *Corporate Law and Accountability Research Group, Working Paper No. 10*, August 2007, Department of Business Law and Taxation, Monash University, p. 11 ff.
7 John Rintoul, *Esperance: Yesterday and today* (4th edn), Esperance Shire Council, Printed by Scott Four Colour Print Pty Ltd, Perth, 1986, pp. 109 and 215–16.
8 Meredith and Dyster, *Australia in the Global Economy*, pp. 4–5.
9 See especially Lionel Frost, 'Across the Great Divide: The economy of the inland corridor', in Mayne (ed.), *Beyond the Black Stump*, pp. 57–84.
10 *Australians: Historical Statistics*, p. 148.
11 Erik Eklund, 'Managers, Workers and Industrial Welfarism: Management strategies at ER&S and the Sulphide Corporation, Cockle Creek, 1895 to 1929', *Australian Economic History Review*, vol. 37, no. 2, 1997, pp. 137–38; and

F. Carrigan, 'The Imperial struggle for control of the Broken Hill base-metal industry, 1914–1915', in K. Buckley and T. Wheelwright (eds), *Essays in the political economy of Australian capitalism, Vol. 5*, Australian New Zealand Book Company, Sydney, 1983, pp. 164–86.

12 Cochrane, *Industrialisation and Dependence*, p. 11.

13 *Australians: Historical Statistics*, p. 148.

14 Jeremy Mouat, 'The development of the flotation process: Technological change and the genesis of modern mining', *Australian Economic History Review*, vol. 36, no. 1, 1996, pp. 3–31.

15 S. Ville and D. Merrett, 'The Development of Large Scale Enterprise in Australia, 1910–64', *Faculty of Commerce Papers*, University of Wollongong, 2000, pp. 17–18.

16 These paragraphs draw on the *Australian Dictionary of Biography* entries for William Knox (1850–1913), William Orr (1843–1929), Anthony Edwin Bowes Kelly (1852–1930); and Charles Rasp (1846–1907). See also Jenny Camilleri, *In the Broken Hill Paddock: Stories from the past*, published by the author, Broken Hill, c.2006.

17 Ian Hore-Lacy (ed.), *Broken Hill to Mount Isa: The mining odyssey of W.H. Corbould*, Hyland House, Melbourne, 1981, p. 40.

18 B. Kennedy, *Silver, Sin and Six Penny Ale: A social history of Broken Hill*, Melbourne University Press (MUP), Melbourne, 1978, pp. 21–24.

19 John Kerr, *Mount Morgan: Gold, copper and oil*, J.D and R.S. Kerr, Brisbane, 1982, p. 117.

20 D.W. Meining, *On the Margins of the Good Earth: The South Australian wheat frontier, 1869–1884*, South Australian Government Printer, Netley (SA), 1988, pp. 60–61; and R.J.R. Donley, *The Rise of Port Pirie*, published by the author, Port Pirie, 1975, pp. 15–16.

21 The narrow gauge line was completed to Cockburn, near the NSW border, in June 1887, and the Silverton Tramway Company completed the link to Broken Hill in January 1888. See Donley, *The Rise of Port Pirie*, pp. 35–36; and Nancy Robinson, *The Reluctant Harbour: The romance of Pirie*, Nadjuri Australia, Jamestown (SA), 1976, pp. 204–207.

22 Donley, *The Rise of Port Pirie*, pp. 41–47.

23 C.B. Johnstone, 'A History of the Origins, Formations and Development of the BHAS Ltd at Port Pirie, South Australia', PhD thesis, University of Melbourne, 1983, p. 214.

24 Peter Richardson, 'The Origins and Development of the Collins House Group, 1915–1951', *Australian Economic History Review*, vol. 27, no.1, 1987, pp. 16–17.

25 G. Blainey, *Mines in the Spinifex: The story of Mount Isa Mines* (revised edn), Angus & Robertson, Sydney, 1965, p. 115; and Hore-Lacy (ed.), *Broken Hill to Mount Isa*, p. 191.

26 N. Kirkman, *Mount Isa: Oasis in the outback*, James Cook University, Townsville, 1998, pp. 14–18. Mount Isa Mines Limited formally changed its name to MIM Ltd in 1970 so the use of this acronym for an earlier period is potentially misleading. For the sake of brevity, however, I will use the acronym MIM to cover the entire period of the company's history from formation in 1924 to its takeover by Xtrata in 2003.

27 Western Mining Corporation Limited formally changed its name to WMC Resources Ltd in 1996, though for the sake of brevity the acronym WMC will

be used to cover the entire period of the company's history from 1933 to its takeover by BHP in 2005.
28. Lindesay Clark, *Built on Gold: Recollections of Western Mining*, Hill of Content, Melbourne, 1983, pp. 1–17; and Don Reid, *They Searched: A history of exploration within Western Mining Corporation, Part 1 Sawdust and Ice: The First Twenty Years*, Western Mining Corporation, Melbourne, 1981, pp. 2–5.
29. Transcript of an interview with Paul Said (by Barry York, 12 September 1984), TRC 3582/2, NLA.
30. Interview with Max Thompson (by John Merritt, 5 December 2001), TRC 4822, NLA.
31. See, for example, Christine Adams, *Sharing the Lode: The Broken Hill migrant story*, Broken Hill Migrant Heritage Committee, Broken Hill, 2004; Interview with Sigge Lampen, p. 6; and Interview with Michael Jones, p. 5.
32. Interview with Margaret Jones, p. 2.
33. Interview with Sigge Lampen, p. 6.
34. Interview with Elizabeth Firth, pp. 1–2.
35. Interview with Sigge Lampen, p. 8.
36. For insights see Janet McCalman, 'Class and Respectability in a Working-Class Suburb: Richmond, Victoria before the Great War', *Historical Studies*, vol. 20, no. 78, 1982, pp. 90–103; and S. Macintyre, *Oxford History of Australia, Volume 4: The Succeeding Age, 1901–1942*, OUP, Melbourne, 1986, pp. 48ff.
37. Macintyre, *Oxford History of Australia*, p. 43
38. B. Penrose, 'Occupational lead poisoning at Mount Isa Mines in the 1930s', *Labour History*, no. 73, November 1997, p. 123.
39. Kennedy, *Silver, Sin and Six Penny Ale*, pp. 119 and 129.
40. Interview with Elizabeth Firth, p. 9.

2 Broken Hill: Icon of working-class culture

1. George Dale, *The Industrial History of Broken Hill*, Fraser & Jenkinson, Melbourne, 1918; Roy Bridges, *From Silver to Steel: The romance of the Broken Hill Proprietary*, Robertson, Melbourne, 1920; Geoffrey Blainey, *The Rise of Broken Hill*, Macmillan, Melbourne, 1968 and *The Rush that Never Ended*, MUP, Melbourne, 1964; Donald McLean, *Broken Hill Sketchbook*, illustrations by Frank Beck, Rigby, Adelaide, 1968; Kennedy, *Silver, Sin and Sixpenny Ale*; R.J. Solomon, *The Richest Lode: Broken Hill, 1883–1988*, Hale & Iremonger, Sydney, 1988; and Paul Robert Adams, *The Best Hated Man in Australia: The life and death of Percy Brookfield, 1875–1921*, Puncher & Wattmann, Sydney, 2010. A sample of the local and family history publications includes; Broken Hill Historical Society, *Newsletter*, 1966–1971; E.F. Woodman, *The Catholic Church in Broken Hill, 1883–1983: The first hundred years*, E.F. Woodman, Broken Hill, 1983; R.H.B. Kearns, *Broken Hill, Volume 4: The first century, 1940–1983* (revised edn), Broken Hill Historical Society, Broken Hill, 1987; and Jenny Camilleri, *Some Outstanding Women of the Broken Hill and District*, Jenny Camilleri, Broken Hill, 2002.
2. Ion Idriess, *The Silver City*, Angus & Robertson, Sydney, 1956; Arthur Upfield, *The Bachelors of Broken Hill*, Heinemann, London, 1958; and Kenneth Cook, *Wake in Fright*, St. Martin's Press, New York, 1962.
3. Bradon Ellem and John Shields, 'Making a "Union Town": Class, Gender and Consumption in Inter-War Broken Hill', *Labour History*, no. 78, 2000, p. 117. More recent work has begun to extend the focus into the 1920s and beyond.

See, for example, Sarah Gregson, 'Defending Internationalism in Interwar Broken Hill', *Labour History*, no. 86, 2004, pp. 115–36; and Julie Kimber, 'A Case of Mild Anarchy?: Job Committees in the Broken Hill Mines, c.1930 to c.1954', *Labour History*, no. 80, 2001, pp. 41–64.

4   Paul McGuire, *Australian Journey* (new and revised edn), William Heinemann, London, 1942, p. 335.

5   Charles Sturt, *Narrative of An Expedition performed under the authority of Her Majesty's Government, during the years 1844, 5, and 6: together with a notice of the province of South Australia, in 1847, Volume 2*, T. & W. Bone, London, 1849, p. 129.

6   See, for example, John Clark (ed.), *Australia in Maps: Great maps in Australia's history from the National Library's Collection*, National Library of Australia, Canberra, 2007, p. 84; and *Australian Encyclopedia Volume 2* (4th edn), Grolier Society, Sydney, 1983, p. 108.

7   Both Sturt's diaries from 1844 and his published work on his expedition reveal that he made reference to the 'Barrier Range' or 'Stanley's Barrier Ranges', but did not refer to a 'Broken Hill'. See Sturt, *Narrative of an Expedition into Central Australia*, University of Adelaide Library Electronic Facsimile Edition, 2008; and Richard Davis (ed.), *The Central Australian Expedition, 1844–1846: The journals of Charles Sturt*, Hakluyt Society, London, 2002. Reference to the 'Broken Hill paddock' comes from *Burra Record*, 3 February 1885, p. 4 and *The Argus*, 15 August 1888, p. 19.

8   Timothy Coghlan, *General Report on the Eleventh Census of New South Wales*, D. Potter, Government Printer, Sydney, 1894, p. 122.

9   Jenny Camilleri, *In the Broken Hill Paddock*, Jenny Camilleri, Broken Hill, 2006, pp. 72–84; and Jeannette Hope, 'Unincorporated Area of NSW – A Heritage Study: A Report for the Department of Natural Resources and the Heritage Office of NSW', River Junction Research, 2006.

10 Interview with May Watson, p. 1.

11 Interview with Harry Finlay, p. 16.

12 The changing names of the Broken Hill companies present particular problems. For example, Broken Hill South was called the Broken Hill South Silver Mining Company when it was floated in 1885. In 1918 it was formally registered as Broken Hill South Limited. During this period and thereafter other leases were acquired and subsidiary companies set up along the line of lode. I have not covered these complex changes to the names and operations of Broken Hill South or the other mining companies; readers are referred to other more detailed mining history sources.

13 J. Lamb, cited in *The Register*, 28 January 1916, p. 5. Laurence Finn was the licensee for the Royal Hotel in Argent Street from 1886 until his death in 1907, apart from a seven-year period from 1891 when it was leased to the South Australian Brewing Company. See *Barrier Miner*, 27 November 1907, p. 2.

14 A.W.B. Orman, '1st Town Building Opened 50 Years Ago To-day', *Barrier Miner*, 31 July 1935, p. 2.

15 Horace Webber, *The Greening of the Hill: Revegetation around Broken Hill in the 1930s*, Hyland House, Melbourne, 1992, pp. 10–19.

16 Kearns, *Broken Hill, Volume 4*, p. 65.

17 Interview with Frank Bartley (by Edward Stokes in the Broken Hill social history project), Oral TRC 1873/39–50, NLA.

18 Dale, *The Industrial History of Broken Hill*, p. 269.

19 *The Silver-Lead Mines of Broken Hill NSW – Death and disease*, Broken Hill, Broken Hill Branch Communist Party, 1933?, p. 1.
20 *Barrier Miner*, 11 February, 1919, p. 4.
21 Kennedy, *Silver, Sin and Six Penny Ale*, pp. 158–74.
22 Adams, *The Best Hated Man in Australia*.
23 See especially Kennedy, *Silver, Sin and Six Penny Ale*, pp. 8–20; *The South Australian Advertiser*, 31 July 1885, p. 7.
24 Interview with Frank Bartley (by Edward Stokes in the Broken Hill social history project), Oral TRC 1873/39–50, NLA.
25 *Barrier Miner*, 8 April 1942, p. 4.
26 Bobbie Hardy, *Water Carts to Pipelines: The history of the Broken Hill Water Supply*, Broken Hill Water Board/Halstead Press, Sydney, 1968, pp. 1–21.
27 Ken McCarthy, *Steaming Down Argent Street: A history of the Broken Hill steam tramways 1902–1926*, The Sydney Tramway Museum, Sydney, 1983.
28 Woodman, *The Catholic Church in Broken Hill*, pp. 3–15.
29 Ibid., pp. 67–73, 89.
30 *Barrier Miner*, 22 January 1934, p. 2. The total population of male and female 'breadwinners' in 1933 was 11,785. See *Census of the Commonwealth of Australia*. 3rd–4th April 1933, Part I. 'New South Wales. Population Detailed Tables for Local Government Areas', pp. 111, 119. On friendly societies generally, see Dan Weinbren and Bob James, 'Getting a Grip: The role of friendly societies in Australia and Britain reappraised', *Labour History*, no. 88, May 2005, pp. 87–103.
31 Kennedy, *Silver, Sin and Six Penny Ale*, pp. 19–20.
32 There is an extensive literature on this area but a few indicative works would include: Stuart Svensen, *Industrial War: The great strikes, 1890–1894*, Ram Press, Wollongong (NSW), 1995; John Rickard, *Class and Politics: New South Wales, Victoria and the Early Commonwealth, 1890–1910*, ANU Press, Canberra, 1976; and R.W. Connell and T.H. Irving, *Class Structure in Australian History: Poverty and progress*, Longman Cheshire, Melbourne, 1992.
33 Graeme Osborne, 'Town and Company', in John Iremonger, John Merritt and Graeme Osborne (eds), *Strikes: Studies in twentieth century social history*, Angus & Robertson in association with the Australian Society for the Study of Labour History, Sydney, 1973, p. 29.
34 Paul Robert Adams and Erik Eklund, 'Representing Militancy: Photographs of the Broken Hill industrial disputes, 1908–1920', *Labour History*, no. 101, 2011, pp. 1–34.
35 Dale, *The Industrial History of Broken Hill*, pp. 247–61.
36 See Bradon Ellem and John Shields, 'Why do Unions form Peak Bodies? The Case of the Barrier Industrial Council', *Journal of Industrial Relations*, vol. 38, no. 3, 1996, pp. 377–411. The Barrier Industrial Council emerged from a newly revived Trades and Labour Council in 1923–24. The Trades and Labour Council changed its name to the Barrier Industrial and Political Council in 1924 and then dropped the 'Political' at the end of 1924.
37 Kennedy, *Silver, Sin and Six Penny Ale*, pp. 16–17.
38 *Barrier Miner*, 2 July 1891, p. 2.
39 Bede Nairn, 'Cann, John Henry (1860–1940)', *Australian Dictionary of Biography*, National Centre of Biography, Australian National University, http://adb.anu.edu.au/biography/cann-john-henry-5497/text9351. Former AMA President

Josiah Thomas, the Labor member for Alma from 1894 to 1901, had a very similar Methodist background. See Bruce Pennay, 'Thomas, Josiah (1863–1933)', *Australian Dictionary of Biography*, National Centre of Biography, Australian National University, http://adb.anu.edu.au/biography/thomas-josiah-8779/text15391.
40   Woodman, *The Catholic Church in Broken Hill, 1883–1983*.
41   Bradon Ellem, John Shields and Julie Kimber, 'The Far West', in Jim Hagan (ed.), *People and Politics in Regional New South Wales, Volume 2: The 1950s to 2006*, Federation Press, Sydney, 2006, p. 293.
42   Ellem, Shields and Kimber, 'The Far West', pp. 282–316.
43   *Barrier Miner*, 11 April 1950, p. 6 and 4 February 1953, p. 2.
44   Kennedy, *Silver, Sin and Six Penny Ale*, pp. 58–59.
45   *The Advertiser*, 24 January 1898, p. 5.
46   Woodman, *The Catholic Church in Broken Hill*, p. 16.
47   *Barrier Miner*, 12 January 1901, p. 4.
48   Graham Burke, *Football Heroes of the Hill: Those wild old westies*, Graham Burke, Melbourne, 1999.
49   Interview with Peter Bennett, p. 11.
50   Peter Lines (compiler) *Encyclopaedia of South Australian Country Football Clubs*, published by the author, Cowell (SA), 2008. I am also indebted to the work of local football historian John Devaney. See his 'A Brief History of Broken Hill Football', http://www.fullpointsfooty.net/a_brief_history_of_football_in_broken_hill.htm.
51   A.W.B. Orman, '1st Town Building Opened 50 Years Ago To-day', *Barrier Miner*, 31 July 1935, p. 2.
52   *Sydney Morning Herald*, 5 May 1898, p. 7.
53   Regional Histories: Regional Histories of New South Wales, NSW Heritage Office, Sydney, 1996, 'Regional Histories – part 7', p. 193.
54   Jeremy Beckett, 'Marginal Men: A study of two half-caste [sic] Aborigines', *Oceania*, vol. 29, no. 2, 1958, pp. 91–108.
55   Beverley Elphick and Don Elphick (compilers and editors), *Menindee Mission Station, 1933–1949*, B. & D. Elphick, Canberra, 1996.
56   *Barrier Miner*, 6 April 1951, p. 8 and 22 October 1954, p. 1.
57   *The Advertiser* (Adelaide), 30 June 1951, p. 4.
58   See, for example, the experiences of one Maltese migrant, Paul Said. Transcript of an interview with Paul Said (by Barry York, 12 September 1984), Oral TRC 3582/2, NLA.
59   See, for example, Adams, *Sharing the Lode: The Broken Hill migrant story*.
60   Interview with Harry Finlay, p. 20.
61   Sarah Gregson, 'Defending Internationalism in Interwar Broken Hill', pp. 115–36.
62   Interview with Clem Davison (by Edward Stokes in the Broken Hill social history project, 8–9 August 1982), Oral TRC 1873/94–95, NLA.
63   *Official Year Book of the Commonwealth of Australia*, Issue no. 46, 1960, p. 1064.
64   *Consolidated Zinc Corporation Limited*, Consolidated Zinc Corporation, London, 1960, 'Ore Flow', no page numbers.
65   *Official Year Book of the Commonwealth of Australia*, Issue no. 40, 1954, p. 768.
66   Kearns, *Broken Hill, Volume 4*, pp. 13–14.
67   Interview with Peter Bennett, p. 9.

68 Interview with Elizabeth Firth, p. 7.
69 Kearns, *Broken Hill, Volume 4*, p. 20.
70 George Farwell, *Down Argent Street: The story of Broken Hill* (illustrated with drawings by Roy Dalgarno), F.H. Johnston Publishing Co. P/L, Sydney, 1948, p. 84.
71 *Consolidated Zinc Corporation Limited*, Consolidated Zinc Corporation, London, 1960, 'Welfare and Training in Australia', no page numbers.
72 Interview with Carole Michaels, p. 10.
73 Lorenzo Cester's story, cited in Adams, *Sharing the Lode: The Broken Hill Migrant Story*, pp. 240–46.
74 Horace Webber, *The Greening of the Hill: Re-vegetation around Broken Hill in the 1930s*, Hyland House, Melbourne, 1992.
75 *Census of the Commonwealth of Australia*, 30 June 1961, vol. 1 'New South Wales, Analysis of Population in Local Government Areas and in Non-Municipal Towns of 1,000 persons or more', pp. 38, 69.
76 *Census of the Commonwealth of Australia*, 30 June 1961, Volume 8, Part 1 'Cross-Classifications of the Characteristics of the Population', p. 27.
77 *A Tale of Tin and Silver: Broken Hill Heritage Study*, Latana, Masterman & Associates, Sydney, 1987.
78 *Census of the Commonwealth of Australia*, 3–4 April 1921, Part XVIII 'New South Wales. Dwellings in Local Government Areas', pp. 1505, 1513.
79 McGuire, *Australian Journey*, p. 303; and Elspeth Huxley, *Their Shining Eldorado: A journey through Australia*, Chatto & Windus, London, 1967, p. 313.
80 Ellem, Shields and Kimber, 'The Far West', p. 317 and Camilleri, *Some Outstanding Women of Broken Hill and District*, pp. 70–71.
81 W.A. Howard, *Barrier Bulwark: The life and times of Shorty O'Neil*, Willry P/L, Kew (Vic), 1990, pp. 168–69.
82 Interview with Peter Bennett, p. 14.
83 Howard, *Barrier Bulwark*, pp. 125–29.
84 *Sydney Morning Herald*, 5 November 1954, p. 2; and *Barrier Miner*, 5 November 1954, p. 1.
85 Lisa Andersen and Jane Andrew, *Quality of Light, Quality of Life: Professional artists and cultural industries in and around Broken Hill*, Regional Arts NSW, Millers Point (NSW), 2007.
86 *Barrier Miner*, 28 February 1913, p. 5; and Kennedy, *Silver, Sin and Six Penny Ale*, pp. 9–95.
87 For details of these works of fiction and non-fiction see footnotes 1 and 2 above.
88 Interview with Peter Bennett, p. 8.
89 See NSW Government, Office of Liquor Gaming and Racing http://www.olgr.nsw.gov.au/gaming_info_two_up_history.asp. The law was amended in 1992 to allow for two-up in Broken Hill under the control of a state government licence.
90 P.M. Yellowlees and A.V. Kaushik, 'The Broken Hill Psychopathology Project', *Australian and New Zealand Journal of Psychiatry*, vol. 26, no. 2, 1992, pp. 197–207.
91 Jenny Onyx, Melissa Edwards and Paul Bullen, 'The Intersection of Social Capital and Power: An application to rural communities', *Rural Society*, vol. 17, no. 3, 2007, pp. 215–30.
92 *Sydney Morning Herald*, 29 January 1953, p. 2.
93 See, for example, *Barrier Miner*, 7 December 1897, p. 1; *The Advertiser*, 13 April

1908, p. 8; and *Barrier Miner*, 26 August 1902, p. 2. The Broken Hill example appears to contrast strongly with international research, which has found a strong correlation between mining regions and prostitution. See Julianne Laite, 'Historical Perspectives on Industrial Development, Mining and Prostitution', *The History Journal*, vol. 52, issue 3, 2009, pp. 739–62.
94 Bob Bottom, *Behind the Barrier*, Gareth Powell Associates, Hong Kong, 1969, p. 108.
95 Huxley, *Their Shining Eldorado*, pp. 78–79.
96 Bottom, *Behind the Barrier*, p. 20.
97 Kearns, *Broken Hill, Volume 4*, pp. 77–79.
98 Interview with Carole Michaels, p. 17.

3 Mount Morgan: In the thrall of modernity

1 Australian Bureau of Meteorology, 'Climate Statistics for Australian Locations', Walterhall (1896–2011) [1.3km from Mount Morgan] http://www.bom.gov.au/climate/data.
2 A.G. Shanks, 'Fifty Years' History' [1939], reproduced in A.G. Shanks and R.F. Boyle, *The First Hundred Years: Mount Morgan Lodge No. 57, United Grand Lodge of Queensland*, Mount Morgan Lodge No. 57, Mount Morgan, 1988, np.
3 See, for example, *The Observer* (London), 5 February 1888, p. 2; *The Argus* (Melbourne), 19 August 1898, p. 7.
4 W.H. Dick, *The Famous Mount Morgan Gold Mine, Rockhampton, Queensland: Its history, geological formation and prospects*, William Munro, Rockhampton, 1888, p. 17. Perhaps the most infamous case was the Taranganba Gold Mine, near Yeppoon, 'discovered' in 1886. Widely heralded and often compared with Mount Morgan, it proved to be worthless. See J. Kerr, *Mount Morgan: Gold, copper and oil*, J.D and R.S. Kerr, Brisbane, 1982, pp. 67–75.
5 Kerr, *Mount Morgan: Gold, copper and oil*, pp. 58–65; and Mount Morgan Gold Mining Company Limited, Report to Shareholders, 1888, Mitchell Library, Sydney.
6 *Queensland Mining and Milling Practice*, Brisbane, 1901, p. 30.
7 Rees Rutland Jones, *Gold Mining in Central Queensland and the Mount Morgan Mine*, Morning Argus, Rockhampton, 1913, pp. 16–17.
8 See Mount Morgan Gold Mine Company Ltd, Annual Reports, 1899–1906, Mount Morgan Manuscript Collection, MS D15/271.1, Capricornia Collection, CQU.
9 J.W. Long, *The Early History of Mount Morgan, Dawson and Callide Valleys*, [Mount Morgan], 1962, p. 2; and Eklund, *Steel Town*, p. 23.
10 *Commonwealth Year Book*, no.1, 1908, p. 403.
11 J.H. Lundager, *The Early History of Rockhampton*, Rockhampton Morning Bulletin, Rockhampton, 1909, p. 1, 235ff.
12 Jones, *Gold Mining in Central Queensland and the Mount Morgan Mine*, pp. 6–7.
13 This work has focused on who truly discovered the gold on Mount Morgan, and on whether various parties were unfairly relieved of their rightful share. See Kerr, *Mount Morgan: Gold, oil and copper*, pp. 17–36; and Cyril Grabs, *Gold, Black Gold and Intrigue: The story of Mount Morgan* (2nd edn), Central Queensland University (CQU) Press, Rockhampton, 2000, pp. 14–40.
14 R. Newtown, *The Work and Wealth of Queensland*, Outride & Co., Brisbane, 1897, p. 44.

Notes to pages 74–82

15 L. Huf, L. McDonald and D. Myers (eds), *Sin, Sweat and Sorrow: The making of Capricornia, Queensland, 1840–1940s*, UQP, Rockhampton, 1993, pp. 15–32, 95–120.
16 See biographical information on various Rockhampton politicians, http://www.parliament.qld.gov.au/communityengagement/view/exhibitions/documents/05CQParl_DisplayPanels.pdf.
17 *Australian Dictionary of Biography*, entry for John Ferguson (1830–1906).
18 *Morning Bulletin*, 28 March, 1887, p. 5; and George Westacott, cited in N. Chardon and F.L. Golding (eds), *Centenary of the Town of Mount Morgan, 1882–1982*, Mount Morgan and District Historical Society, Mount Morgan, 1982, p. 9.
19 *Morning Bulletin*, 18 March 1889, p. 5.
20 N. Chardon and F.L. Golding (eds), *Centenary of the Town of Mount Morgan, 1882–1982*, Mount Morgan and District Historical Society, Mount Morgan, 1982, pp. 35–37.
21 *Morning Bulletin*, 14 July 1887, p. 5.
22 *Morning Bulletin*, 21 May 1889, p. 5.
23 *Morning Bulletin*, 4 October 1887, p. 5.
24 B. Cosgrove, 'Mount Morgan: Images and realities – dynamics and decline of a mining town', PhD thesis, CQU, 2001, p. 65.
25 Recollections of Mrs C.H. Brock published in *Morning Bulletin*, 14 June 1927, p. 7.
26 Cosgrove, 'Mount Morgan: Images and realities', pp. 67–70.
27 *Morning Bulletin*, 31 August, 1889, p. 5.
28 Chardon and Golding (eds), *Centenary of the Town of Mount Morgan, 1882–1982*, p. 13.
29 *Morning Bulletin*, 30 April 1888, p. 4; *Morning Bulletin*, 15 August 1906, p. 3; and Chardon and Golding (eds), *Centenary of the Town of Mount Morgan, 1882–1982*, p. 53. On the Hibernians and other fraternal orders see Cosgrove, 'Mount Morgan: Images and Realities', pp. 166–97.
30 The Queensland average is calculated from the 1921 Census. See *Census of the Commonwealth of Australia*, 3–4 April, 1921, Part XII – Queensland, 'Population of Local Government Areas', pp. 868–73.
31 *The Queenslander*, 31 December 1892, p. 1277; *Australian Dictionary of Biography*, entry for Brice Frederick Bunny (1820–1885); *Morning Bulletin*, 3 August 1905, p. 5.
32 *Morning Bulletin*, 3 September 1889, p. 3.
33 Cosgrove, 'Mount Morgan: Images and realities', pp. 67–70.
34 *Morning Bulletin*, 30 August 1892, p. 4.
35 Editorial, *Morning Bulletin*, 28 September 1893, p. 2. On the conservatism of the *Morning Bulletin* see Cosgrove, 'Mount Morgan: Images and realities', p. 235.
36 There is an extensive literature on this area with some major contributions cited in Chapter 2 footnote 32. On Queensland's first (albeit short-lived) Labor government in 1899, under Anderson Dawson, see D.J. Murphy, 'The Dawson Government in Queensland, The First Labour Government in the World', *Labour History*, no. 21, 1971, pp. 1–8.
37 *Morning Bulletin*, 21 February 1899, p. 9 and 27 March 1899, p. 6 ; and Kerr, *Mount Morgan: Gold, copper and oil*, p. 116.
38 L. McDonald, *Rockhampton: A history of city and district*, UQP, Brisbane, 1980, pp. 314, 330–32.

39 Henri Cowap, cited in the *Morning Bulletin*, 13 February 1902, pp. 2–3. Cowap held the seat of Fitzroy until 1909.
40 See L. McGerity, '"White Queensland": The Queensland Government's Ideological Position on the Use of Pacific Island Labourers in the Sugar Sector 1880–1901', *Australian Journal of Politics and History*, vol. 52, no. 1, 2006, pp. 1–12.
41 Kerr, *Mount Morgan: Gold, copper and oil*, p. 29; transcript of interview with Arthur Barnham (by B. Cosgrove, 23 April 1993), Capricornia Collection, CQU, pp. 1–2; and transcript of interview with Bill Toby (by B. Cosgrove, 29 April and 5 May 1993), Capricornia Collection, CQU, pp. 1–11.
42 Transcript of interview with Arthur Barnham (by B. Cosgrove, 23 April 1993), Capricornia Collection, CQU, pp. 1–2; transcript of interview with Bill Toby (by B. Cosgrove, 29 April and 5 May 1993), Capricornia Collection, CQU, pp. 1–11; and Palmtree Wutaru Aboriginal Corporation for Land and Culure (in association with Central Queensland Cultural Heritage Management), 'Report on the Aboriginal Cultural Values of the Area to be affected by the Proposed Redevelopment of the No. 7 Dam, Mt Morgan, Queensland', prepared for Mcintyre and Associates, Brisbane, October, 1997, p. 7.
43 William Gordon Cowie, *Some Account of the Mount Morgan Gold Mine, Rockhampton, Queensland*, Morning Bulletin, Rockhampton, 1885; and W.H. Dick, *A Mountain of Gold: Origin, history, geological features, latest developments, etc., of the famous Mount Morgan mine: with a complete description of the chlorination process and new machinery: and estimates of value and permanence*, Woodcock & Powell, Brisbane, 1889.
44 Hume Nisbet, *A Colonial Tramp: Travels and adventures in Australia and New Guinea*, Ward & Downey, 1891, p. 79.
45 R. Newtown, *The Work and Wealth of Queensland*, Outride & Co., Brisbane, 1897, p. 41.
46 Ibid., p. 44.
47 Ville and Merrett, 'The Development of Large Scale Enterprise in Australia, 1910–64', p. 33.
48 *The Story of the Mount Morgan Mine, 1882–1957*, Mount Morgan Ltd, Mount Morgan, 1957, pp. 12–13; and Kerr, *Mount Morgan: Gold, copper and oil*, p. 156.
49 Eklund, *Steel Town*, pp. 39–40.
50 C. Gistitin and R.F. Boyle, 'The Golden Mount – Central Queensland's First Triumph', *Proceedings of the Fifth National Conference on Engineering Heritage*, 3–5 December 1990, Perth, pp. 100–01.
51 See entry for James (Jimmy) Stopford (1878–1936), union official from Mount Morgan and later Labor politician and Minister for Mines, http://www.adbonline.anu.edu.au/biogs/A120122b.html; J. Hagan, *A History of the ACTU*, Longman Cheshire, Melbourne, 1981, p. 13; and Mark Hearn and Harry Knowles, *One Big Union: A history of the Australian Workers Union, 1886–1994*, CUP, Cambridge, 1996.
52 *Census of the Commonwealth of Australia*, 3–4 April 1921, Vol. XII Queensland, 'Population in Local Government Areas', pp. 892–901.
53 See G. Griffin, 'J.H.Lundager, Mount Morgan Politician and Photographer: Company Hack or Subtle Subversive?', *Journal of Australian Studies*, no. 34, September 1992, pp. 15–31.
54 Turner, *Industrial Labour and Politics*, p. 194.
55 *Daily Mail*, 13 April 1921.
56 *The Argus*, 23 May 1921 and 6 July 1921; *The Age*, 7 July 1921.

57 *The Argus*, 21 February 1922, p. 4 (editorial).
58 See McDonald, *Rockhampton: A history of city and district*, pp. 318–21.
59 Transcript of interview with Colin and Vi Heberlein (by B. Cosgrove, 21 October 1992), Capricornia Collection, CQU, p. 2.
60 See, for example, Adam Boyd, 'A History of Mount Morgan', *Australasian Institute of Mining and Metallurgy, Proceedings, New Series*, no. 115, Melbourne, 1939, p. 264; and Mount Morgan Gold Mining Company Limited, Directors' Report to Shareholders, 1925, p. 5, Mitchell Library, Sydney.
61 Interview with Arthur Timms (by Ray Aitchison, 14 August 1986), Oral TRC 2061, NLA.
62 *Rockhampton Morning Bulletin*, 12 July 1929.
63 Transcript of interview with Bill Butcher (by Betty Cosgrove, 16 October 1992), Capricornia Collection, CQU, p. 4.
64 Transcript of interview with Charlie Shannon (by Betty Cosgrove, 10 September 1992), Capricornia Collection, CQU, p. 2.
65 Interview with Jim Leigh (by Betty Cosgrove, 15 September 1992), Capricornia Collection, CQU, p. 4.
66 *Evening News*, 28 December 1934 and Mount Morgan Ltd, Report and Statement of Account for the Year ending 1935. On the welfare policy details and the funding commitment see 'Minutes of the Special Meeting of the Works Committee', 12 February 1934, Mount Morgan Manuscript Collection, MS D14/1281.1, Capricornia Collection, CQU.
67 Transcript of interview with Taffy Badham (by Carol Gistitin, 5 August 1989), Capricornia Collection, CQU, p. 2. See also McDonald, *Rockhampton: A history of city and district*, pp. 326–27.
68 Interview with Patrick O'Regan, p. 11.
69 *Census of the Commonwealth of Australia, 1954*, Vol. 3, 'Analysis of Population in Local Government Areas', pp. 140–41.
70 *Census of the Commonwealth of Australia, 1954*, Vol. 3, 'Analysis of Dwellings in Local Government Areas', pp. 16, 28.
71 *Operations of Mount Morgan Limited, Mount Morgan, Queensland* (7th edn), Mount Morgan, 1969, p. 27.
72 *Quin's Metal Handbook and Statistics*, Metal Information Bureau, London, 1953; F.E. Banks, *Copper in the World Economy*, Ballinger Publishing Company, Cambridge (MA), 1974, p. 12; *Official Year Book of the Commonwealth of Australia*, no. 44, 1955, p. 764.
73 *Official Year Book of the Commonwealth of Australia*, 1960, pp. 1052–53.
74 *Official Year Book of the Commonwealth of Australia*, 1970, pp. 910–14.
75 Transcript of interview with Sydney (Syd) Wall (by B. Cosgrove, 10 November 1992), Capricornia Collection, CQU; and interview with Bryce Hargreaves (by B. Cosgrove, 9 November 1992), Capricornia Collection, CQU, pp. 12–13.
76 Transcript of interview with Lorna McDonald (by B. Cosgrove, 10 and 19 August 1992), Capricornia Collection, CQU, p. 4.
77 Transcript of interview with Charlie Shannon (by B. Cosgrove, 10 September 1992), Capricornia Collection, CQU, p. 8.
78 Transcript of interview with Bryce Hargreaves (by B. Cosgrove, 9 November 1992), Capricornia Collection, CQU, pp. 12–13.
79 Ibid., p. 4.
80 Transcript of interview with Christine Schluter (by B. Cosgrove, 5 November

1992), Capricornia Collection, CQU pp. 4–5.
81 Gistitin and Boyle, 'The Golden Mount – Central Queensland's First Triumph', p. 101.
82 Transcript of interview with Donny McDonald (by B. Cosgrove, 16 October 1992), Capricornia Collection, CQU, p. 5.
83 Transcript of interview with Alex Teeny (by B. Cosgrove, 8 October 1992), Capricornia Collection, CQU, p. 2.
84 *Census of the Commonwealth of Australia, 30th June 1966*, Vol. 3, 'Population and Dwellings in Local Government Areas', Queensland, Part 3, pp. 90–91. The total population given here covers the 'urban' parts of the Shire of Mount Morgan. Elsewhere the 1966 Census gives a town population of 4055.
85 Interview with Patrick O'Regan, pp. 6–9; *Census of the Commonwealth of Australia, 30th June 1966*, Vol. 3, 'Population and Dwellings in Local Government Areas', Queensland, Part 3, p. 90.
86 Interview with Arthur Timms (by Ray Aitchison, 14 August 1986), Oral TRC 2061, NLA.
87 Interview with George Dial, p. 12.
88 J. Malcolm Newman, *Guide to the Operations of Mount Morgan Limited*, Mount Morgan, 1960, p. 1.
89 *Evening News*, 4 April 1934.
90 *Morning Bulletin*, 12 December 1935.
91 Kerr, *Mount Morgan: Gold, copper and oil*, p. 156.
92 Adam Boyd, 'A History of Mount Morgan', pp. 267–68. See also McDonald, *Rockhampton: A history of city and district*, pp. 314, 322–23.
93 Interview with Patrick O'Regan, pp. 9–10. See also transcript of interview with Alex Teeny (by B. Cosgrove, 8 October 1992), Capricornia Collection, CQU, p. 2.
94 *Australian Commodity Statistics, 1995*, ABARE, Canberra, 1995; Correspondence from the London Metal Exchange to Erik Eklund, October 1999.
95 Transcript of interview with Bryce Hargreaves, p. 8; and R.F. Boyle and C. Gistitin, 'Mount Morgan – A New Beginning', *Proceedings of the Sixth National Conference on Engineering Heritage*, 5–7 October 1992, Hobart, p. 128.
96 *Morning Bulletin*, 1 June 1984, p. 1.
97 *Census of the Commonwealth of Australia, 30th June 1961*, Vol. III, Part V, 'Queensland: Population and Dwellings in Localities & Census of Population and Housing', p. 27; and *Census of the Commonwealth of Australia, 30 June 1981*, 'Persons and Dwellings in Local Government areas and Urban Centres, Queensland', p. 35.
98 Transcript of interview with George Beattie (by Lorna McDonald, 20 November 1992), Capricornia Collection, CQU, pp. 1–2.
99 Chardon and Golding (eds), *Centenary of the Town of Mount Morgan, 1882–1982*, p. 7.
100 Transcript of interview with Trevor James Stock (by Lorna McDonald, 24 November 1992), Capricornia Collection, CQU, p. 1.
101 Census, 2006, National Regional Profile, Rockhampton, Mount Morgan Statistical Local Area, http://www.abs.gov.au/AUSSTATs/abd@.nsf/781eb7868 cee03e9ca257180008bece/9a422 16f056cfd8dca25771300182884!OpenDocument.
102 Palmtree Wutaru Aboriginal Corporation for Land and Culture (in association with Central Queensland Cultural Heritage Management), 'Report on

the Aboriginal Cultural Values of the Area to be affected by the Proposed Redevelopment of the No. 7 Dam, Mt Morgan, Queensland', pp. 6–7.
103 Interview with Charles Copeman (by John Farquharson, 8 and 10 February 2000), Oral TRC 3976, NLA.

4 Queenstown: 'They've got to come here and they've got to learn about it'

1. Henry Reynolds, *Fate of a Free People*, Penguin, Melbourne, 1995; Kay Daniels, *Convict Women*, Allen & Unwin, Sydney, 1998; James Boyce, *Van Diemen's Land*, Black Inc, Melbourne, 2008; and Hamish Maxwell-Stewart, *Closing Hell's Gates: The death of a convict station*, Allen & Unwin, Sydney, 2008. Likewise, the recent work by Henry Reynolds, *A History of Tasmania* (CUP, Melbourne, 2011) is heavily weighted towards the 19th century.
2. See especially the work published in the *Bulletin of the Centre for Tasmanian Historical Studies* (1985–2011).
3. Geoffrey Blainey, *The Peaks of Lyell* (5th edn), St David's Park Publishing, Hobart, 1993; Charlie Fox, 'Work and Welfare at Mount Lyell, 1913–1923', *Journal of Australasian Mining History*, vol. 1, no.1, 2003, pp. 90–110 and 'Contract Work at Mount Lyell', *Journal of Australasian Mining History*, vol. 6, September 2008, pp. 90–110; and William J. Lines, *Taming the Great South Land: A history of the conquest of nature in Australia*, University of California Press, Berkeley, 1991.
4. Interview with Paul Richardson, p. 13.
5. Interview with Janice Redman, p. 6.
6. *Launceston Examiner*, 22 March 1897, p. 6.
7. *Launceston Examiner*, 2 January 1886, p. 3.
8. Blainey, *The Peaks of Lyell*, pp. 35–41.
9. Ibid., pp. 48–49.
10. Ibid., pp. 42–62.
11. Ibid., pp. 63–68; and S. Ville and D. Merrett, 'The Development of Large Scale Enterprise in Australia, 1910–64', *Business History*, vol. 42, no. 3, 2000, p. 33.
12. Australian Bureau of Meteorology, 'Climate Statistics for Australian Locations', Queenstown (1964–95), www.bom.au/climate/data.
13. A.G. Skuja, 'Forest Resources, Their Development and Effects', in M.R. Banks amd J.B. Kirkpatrick (eds), *Landscape and Man: The interaction between man and environment in Western Tasmania. The Proceedings of a Symposium Organized by the Royal Society of Tasmania*, Royal Society of Tasmania, Hobart, June 1977, pp. 113–18.
14. Population figures for 1891 are from the Tasmanian Colonial Census, *ANU Historical Census and Data Archive*, hccda.anu.edu.au/documents/TAS-1981-census.
15. Blainey, *The Peaks of Lyell*, pp. 50–56.
16. Trevor R. Lee, 'The Social and Demographic Structure of Tasmania's West Coast', in Banks and Kirkpatrick (eds), *Landscape and Man: The interaction between man and environment in Western Tasmania*, pp. 101–11.
17. See *Mount Lyell Mining Lease Cultural Heritage Assessment Study, Vol. 1 Main Report*, report prepared for Mineral Resources Tasmania and the Department of Environment and Land Management, May 1994, Godden Mackay Pty Ltd, p. ii.
18. Geoffrey Blainey, 'Crotty, James (1845–1898)', *Australian Dictionary of Biography*, National Centre of Biography, Australian National University, www.adb.anu.edu.au/biography/crotty-james-5830/text9901.

19 Copy of *Tasmanian Mail*, 30 June 1896 in *Mount Lyell Mining and Railway Company Records*, TSA, NS 1711/934 Volume of newspaper cuttings 30 June 1896–21 December 1900.
20 Robert Sticht, 'The Development of Pyritic Smelting: An outline of its history', Presidential address published in the *Proceedings of the Australian Institute of Mining Engineers*, vol. II, no. II, 1905, Arnell & Jackson/Australian Institute of Mining Engineers, Melbourne, 1905, pp. 9–78.
21 Svenja Keele, 'Lunar Landscapes and Sulphurous Smogs: An environmental history of human impacts on the Queenstown region in western Tasmania c. 1890–2003', *Research Paper No. 18*, School of Anthropology, Geography and Environmental Studies, University of Melbourne, 2003, pp. 9–10.
22 *Launceston Examiner*, 12 November 1895, p. 3.
23 *The Mercury*, 17 December 1896, p. 3. All references to this paper in this chapter refer to the Hobart-based newspaper.
24 *The Mercury*, 22 February 1897, p. 3.
25 *Launceston Examiner*, 7 August 1897, p. 7.
26 *The Mercury*, 29 January 1897, p. 3.
27 T.A. Coghlan, *A Statistical Account of Australia and New Zealand, 1903–1904*, Commonwealth and NSW governments, Sydney, 1904, p. 152 and Alison Alexander, 'Queenstown', *The Companion to Tasmanian History*, www.utas.edu.au/library/companion_to_tasmanian_history/Q/Queenstown.htm.
28 *St. Andrew's, 'the friendly church'*, Queenstown, Tasmania, St Andrews, Queenstown, 1956.
29 *Launceston Examiner*, 7 November 1898, p. 6.
30 *Census of the Commonwealth of Australia, 3rd–4th April, 1921*, Vol. XV Tasmania, 'Population in Local Government Areas', pp. 1124–27.
31 W.T. Southerwood, *Planting a faith in Queenstown*, The Catholic Church, Queenstown, 1974.
32 *Census of the Commonwealth of Australia, 3rd–4th April, 1921*, Vol. XV Tasmania, 'Population in Local Government Areas', pp. 1124–27.
33 A. Cruikshank (secretary of the Hospital Union), cited in *Launceston Examiner*, 26 March 1898, p. 7.
34 *Launceston Examiner*, 30 July 1898, p. 10.
35 *Census of the Commonwealth of Australia, 3rd–4th April, 1921*, Vol. XV Tasmania, 'Population in Local Government Areas', p. 1114.
36 W.A. Townsley, *Tasmania: From colony to statehood*, St David's Park Publishing, Hobart, 1991, p. 289.
37 Charlie Fox, 'Contract Work at Mount Lyell', *Journal of Australasian Mining History*, vol. 6, 2008, pp. 90–110.
38 Ibid., p. 101.
39 See, for example, *The Mercury*, 30 March 1920 p. 4; *Examiner*, 31 August 1928, p. 5; *The Mercury*, 10 October 1939, p. 11.
40 See, for example, *Advocate*, 20 December 1921.
41 Letter from General Manager to W.E. Gardner (Manager, Central Mine Broken Hill) 7 January 1929 in *Mount Lyell Mining and Railway Company Records*, Tasmanian State Archives (TSA), NS 1711/62.
42 'List of Staff Members Employed – 1920 and June 1922', in *Mount Lyell Mining and Railway Company Records*, TSA, NS 1711/62.
43 Tony Weston, 'Mining Lower Grade Ore: Changes in mining technology at

Mount Lyell, Tasmania, 1927–1939', *Journal of Australasian Mining History*, vol. 8, 2010, p. 174.
44 *Census of the Commonwealth of Australia, 3th–4th April 1921*, Vol. XV Tasmania, 'Population in Local Government Areas', p. 1153 and *Census of the Commonwealth of Australia, 30 June 1947*, Part VI Tasmania, 'Analysis of Population in Local Government Areas', p. 458.
45 *Census of the Commonwealth of Australia, 30 June 1933*, Part VI Tasmania – Population, 'Detailed Tables for Local Government Areas', pp. 540–42.
46 Rhys Jones, 'Tasmanian Archaeology: Establishing the sequences', *Annual Review of Anthropology*, vol. 24, 1995, pp. 423–46; and Ian McFarlane, *Beyond Awakening: The Aboriginal tribes of north west Tasmania: A History*, Fullers Bookshop/Riawunna/University of Tasmania, Launceston, 2008, pp. 3–8.
47 Boyce, *Van Diemen's Land*, pp. 296–306.
48 *The Mercury*, 30 September 1922, p. 13.
49 See, for example, Erik Eklund, 'Official and Vernacular Public History: Historical anniversaries and commemorations in Newcastle, NSW', *Public History Review*, vol. 14, 2007, pp. 128–52.
50 *Barrier Miner*, 27 October 1925, p. 2.
51 *The Mercury*, 29 October 1925, p. 6.
52 *Advocate*, 13 April 1926, p. 5.
53 *The Mercury*, 10 December 1925, p. 9.
54 Letter from General Manager to Mount Lyell Company Secretary, Melbourne, 16 April 1926, in *Mount Lyell Mining and Railway Company Records*, TSA, NS 1711/62.
55 *Census of the Commonwealth of Australia, 30 June 1933*, Part VI Tasmania – Population, 'Detailed Tables for Local Government Areas', pp. 532, 534.
56 *Census of the Commonwealth of Australia, 30 June 1947*, Part VI Tasmania, 'Analysis of Population in Local Government Areas', pp. 446, 448.
57 Ian McShane, 'Sticht, Robert Carl (1856–1922)', *Australian Dictionary of Biography*, National Centre of Biography, ANU, www.adb.anu.edu.au/biography/sticht-robert-carl-8670/text15163.
58 Charlie Fox, 'Work and Welfare at Mount Lyell, 1913–1923', *Journal of Australasian Mining History*, vol. 1, no. 1, 2003, p. 56.
59 For more detail on industrial welfarism in the base metals mining and smelting industry see Erik Eklund, 'Intelligently Directed Welfare Work'?: Labour management strategies in local context: Port Pirie, 1915–29, *Labour History*, vol. 76, 1999, pp. 125–48 and 'Managers, workers and industrial welfarism: Management strategies at ER&S, Port Kembla and the Sulphide Corporation, Cockle Creek, 1895–1929', pp. 137–57.
60 'Copper Mining at Mount Lyell', Reprint from the *Chemical Engineering and Mining Review*, November 1940–February 1941, Tait Publishing Company, Melbourne, 1941?, pp. 4–7.
61 These biographical details are from 'The Parliament of Tasmania from 1856', www.parliament.tas.gov.au/history/tasparl/tasparl.htm.
62 Heather Gaunt, 'The Library of Robert Carl Sticht', *La Trobe Journal*, no. 79, 2007, pp. 4–26.
63 *Advocate*, 26 July 1930, p. 10.
64 *The Herald*, 14 and 28 December 1910.
65 *Launceston Examiner*, 17 January 1918, p. 4; and Fox, 'Contract Work at Mount

Notes to pages 123–132

Lyell, 1903–1923', p. 101.
66 Townsley, *Tasmania: From Colony to Statehood*, p. 289.
67 Interview with Paul Richardson, p. 3.
68 See, for example, *The Mercury*, 15 July 1954, p. 18 and 6 September 1952, p. 9.
69 *Census of the Commonwealth of Australia, 30 June 1947*, Part VI 'Analysis of Population in Local Government Areas', Tasmania, pp. 458–61.
70 *Launceston Examiner*, 1 December 1950, p. 16.
71 Ibid., p. 10.
72 *Census of the Commonwealth of Australia, 30 June 1961*, Part VI Tasmania, 'Analysis of Population in Local Government Areas and in non-municipal towns of 750 persons or more', p. 2.
73 Ibid., pp. 10–17.
74 *Census of the Commonwealth of Australia, 30 June 1947*, Part XXV 'Analysis of Dwellings in Local Government Areas – Tasmania', pp. 1915.
75 Interview with Janice Redman, p. 2.
76 *Census of the Commonwealth of Australia, 30 June 1947*, Part XXV 'Analysis of Dwellings in Local Government Areas – Tasmania', pp. 1905–15.
77 Interview with Janice Redman, p. 2. See also interview with Edith Gaynor, pp. 1–3.
78 *Launceston Examiner*, 19 September 1898, p. 4.
79 *Launceston Examiner*, 11 January 1906, p. 11.
80 *Advocate*, 5 October 1926, p. 4.
81 Interview with Dorothy Brown, p. 7.
82 Ibid., p. 6.
83 Interview with Queenstown resident.
84 *Advocate*, 7 January 1938, p. 8.
85 *Examiner*, 17 May 1941, p. 4.
86 *Advocate*, 20 January 1951, p. 15.
87 *The Mercury*, 11 November 1953, p. 25.
88 *Australian Commodity Statistics, 1995*, ABARE, Canberra, 1995 amd Correspondence from the London Metal Exchange to Erik Eklund, October 1999.
89 *Mount Lyell Mining Lease Cultural Heritage Assessment Study, Vol. 1 Main Report*, p. 26.
90 Interview with Janice Redman, pp. 7–8.
91 See Stefan Petrow, 'Saving Tasmania?: The anti-transportation and Franklin River campaigns', *Tasmanian Historical Studies*, vol. 14, 2009, pp. 107–36.
92 Jillian Koshin, 'Eric Reece, 1909–1999', *Companion to Tasmanian History*, www.utas.edu.au/library/companion_to_tasmanian_history/R/Reece.htm.
93 L. Koehnken, *Final Report: Mount Lyell remediation research and demonstration program*, Supervising Scientist Report 126, 1997, p. 3; and P. Davies, N. Mitchell and L. Barmuta, *The impact of historical mining operations at Mount Lyell on the water quality and biological health of the King and Queen River catchments, Western Tasmania*, Supervising Scientist Report 118, 1996, Canberra, p. 3.
94 'Report of the Auditor-General, No. 1 of 1995, Public Account, 1994/1995, September, 1995', *Tasmanian Audit Office*, www.audit.tas.gov.au/publications/reports/report1/1995rep1full.html.
95 'Mount Lyell Mine', Department of Infrastructure, Energy and Resources – Mineral Resources Tasmania, Tasmanian government, www.mrt.tas.gov.au/portal/page?_pageid=35,831249&_dad=portal&_schema=PORTAL.
96 K.D. Corbet, 'New Mapping and Interpretations of the Mount Lyell Mining

District, Tasmania: A Large Hybrid Cu-Au system with an Exhalative Pb-Zn Top', *Economic Geology*, vol. 96, 2001, p. 1089.
97 Lines, *Taming the Great South Land*, p. 44.
98 Keele, 'Lunar Landscapes and Sulphurous Smogs', p. 11.
99 Davies, Mitchell and Barmuta, *The Impact of historical mining operations at Mount Lyell*, p. 1.
100 Ibid.
101 Koehnken, *Final Report: Mount Lyell Remediation Research and Demonstration Program*, 1997.
102 C.V. McQuade, J. Johnston and S.M. Innes, *Review of Historical Literature on the sources and quality of effluent of the Mount Lyell mine site*, Supervising Scientist Report 104, 1995, Canberra, p. 16.

5 Port Pirie: 'Essentially hard and practical'

1   Australian Bureau of Meteorology, 'Climate Statistics for Australian Locations', Port Pirie Nyrstar Smelter (1877–2011) www.bom.gov.au/climate/data/.
2   Nancy Robinson, *The Reluctant Harbour: the romance of Pirie*, Nadjuri Australia, Jamestown (SA), 1976, pp. 29–30; and 'Developing trade and port histories: Outports – Port Pirie', SA Memory – South Australia: Past, Present and Future, www.samemory.sa.gov.au/site/page.cfm?u=682#e1563.
3   *The Australia Directory* (10th edn), Hydrographic Office, British Admiralty.
4   *Commonwealth of Australia Yearbook*, 1908, p. 301.
5   G. Alpin, S. Glynn and M. McKernan (eds), *Australians: Events and places*, Fairfax, Syme & Weldon Associates, Sydney, 1987, p. 354.
6   J. Len Williams, 'The Port Pirie Story: From Settlement to City' (updated and amended by Malcolm C.M. Bottral), Port Pirie, 2007 [1976], p. 5.
7   L.A. Hercus, *A Nukunu Dictionary*, L.A. Hercus, Canberra, 1992.
8   The population figures for 1891 are from the South Australian Colonial Census, *ANU Historical Census and Data Archive*, hccda.anu.edu.au/documents/SA-1891-census.
9   R.J.R. Donley, *The Rise of Port Pirie*, R.J.R. Donley, Adelaide, 1975, pp. 41–47.
10 C.B. Johnstone, 'A History of the Origins, Formations and Development of the BHAS Ltd at Port Pirie, South Australia', PhD thesis, University of Melbourne, 1983, p. 214; Frank Green, *The Port Pirie Smelters*, Broken Hill Associated Smelters, Melbourne, 1977, p. 1; *The Lead and Zinc Industry at Port Pirie, South Australia* (new edn), Broken Hill Associated Smelters, Melbourne, 1982, p. 1.
11 Susan Marsden, 'Playford's Metropolis', in Bernard O'Neil, Judith Raftery and Kerrie Round (eds), *Playford's South Australia: Essays on the history of South Australia, 1933–1968*, Association of Professional Historians, Adelaide, 1996, pp. 117–34.
12 *Census of the Commonwealth of Australia, 30 June 1961*, Vol. IV South Australia, Part I – 'Analysis of Population in Local Government Areas and in Non-municipal Towns of 1,000 Persons or More'.
13 Peter Bell, 'The Heritage of the Upper North: A short history', Professional Historians Association (South Australia), www.sahistorians.org.au/175/documents/the-heritage-of-the-upper-north-a-short-history.shtml, p. 13.
14 Donley, *The Rise of Port Pirie*, p. 13.
15 Robinson, *The Reluctant Harbour*, pp. 191–93.
16 Interview with Leonard Zubrinich (by Edward Stokes in the Broken Hill social history project, 26 June 1982), NLA.

17 For examples of Pirie-based businesses who invested in Silverton and Broken Hill see Robinson, *The Reluctant Harbour*, pp. 182–95.
18 D.W. Meinig, 'A Comparative Historical Geography of Two Railnets: Columbia Basin and South Australia', *Annals of the Association of American Geographers*, vol. 52, no. 4, 1962, pp. 394–413.
19 *Port Pirie Conservation Study*, Lester, Firth and Murton Pty Ltd, Adelaide, 1980.
20 Town Gardener's Annual Report, 1922–1923, in *Port Pirie Mayoral Report, 1922–1923*, p. 17, Mortlock Library, South Australia.
21 Interview with Leonard Zubrinich (by Edward Stokes in the Broken Hill social history project, 26 June 1982), NLA.
22 'Picturesque Port Pirie', *The Critic*, 1 April 1899.
23 *The Recorder*, 2 January 1923, editorial.
24 For the number of convictions for public drunkenness see *Port Pirie Mayoral Report, 1909–1910*, p. 7, Mortlock Library, South Australia.
25 *Census of the Commonwealth of Australia, 3–4 April 1921*, Part XIII 'South Australia, Population of Local Government Areas', pp. 930–31, 940–41. These figures are for the Port Pirie local government area, which includes some rural areas surrounding the town.
26 For the campaign against the ASA see *Port Pirie Recorder*, 13 January and 17 March 1917. For the coverage of the AWU at Port Pirie see Noel Butlin Archives, Canberra, Australian National University (NBAC/ANU), Australian Workers' Union (AWU) Adelaide and South Australia Branch, E154/58, State Secretary's Report for the year ended 31 May 1917, p. 6, Annual Reports and Balance Sheets; and Jim Moss, *The Sound of Trumpets: History of the labour movement in South Australia*, Wakefield Press, Gondawilla (SA), 1985, p. 258.
27 *Port Pirie Recorder*, 9 and 10 August 1917.
28 NBAC/ANU, Waterside Workers' Federation Branch Correspondence, T62/12/4/2, Letter from L.J. O'Malley (Port Pirie branch secretary) to General Secretary, J. Morris, 25 January 1921. Despite the affiliation, the branch was still commonly known as the AWA.
29 NBAC/ANU, Amalgamated Engineering Union, E200/7, Minutes of the Port Pirie Branch of the ASE, 2 July 1915–26 November 1923. By 1922 the first Labor mayor, J.C. Fitzgerald, had been elected to Port Pirie Council.
30 Donley, *The Rise of Port Pirie*, pp. 49–53; and NBAC/ANU, Amalgamated Engineering Union, Minutes of the Federated Society of Iron Shipbuilders and Boilermakers, 28 August 1918–16 May 1928.
31 See, for example, George Dale, *The Industrial History of Broken Hill*, Melbourne, 1918, pp. 130–39, 194–200.
32 Brian Dickey, 'South Australia', in D.J. Murphy (ed.), *Labor in Politics: The state Labor parties in Australia 1880–1920*, UQP, Brisbane, 1975, p. 271.
33 For a sample of reports on the Chamber's activities see *The Register*, 18 November 1910, p. 10; *The Advertiser*, 21 January 1913, p. 10; and *Barrier Miner*, 8 July 1947, p. 7.
34 Donley, *The Rise of Port Pirie*, pp. 19, 27.
35 For Wright's outline of welfarism see C. Wright, *The Management of Labour: A history of Australian employers*, MUP, Melbourne, 1995, pp. 20–24, 33–34.
36 W.S. Robinson, *If I Remember Rightly: The memoirs of W.S. Robinson* (ed. Geoffrey Blainey), Cheshire, Melbourne, 1967, p. 88.
37 Colin Fraser, 'Port Pirie Zinc Smeltery Report', No. 2, 4 February 1916, p. 4,

Sir Colin Fraser Papers, Broken Hill Associated Smelters Proprietary Limited Collection (Colin Fraser Papers), 1/18/6/2, Melbourne University Archives (MUA).
38 *Barrier Daily Truth*, 31 August 1926, editorial.
39 Johnstone, 'A History of the Origins', pp. 320–21.
40 W.S. Robinson to Colin Fraser, 20 January 1917, 'Welfare', Colin Fraser Papers, 1/19/4/2/2, MUA.
41 For a full list of references see Eklund, 'Intelligently Directed Welfare Work?: Labour Management Strategies in Local Context: Port Pirie, 1915–29', p. 131, footnote 25.
42 'Port Pirie Smelters, 1920' by CDW [?], 22 May 1920, p. 21, Colin Fraser Papers, 1/22/3/1, MUA.
43 For the Sickness and Accident Fund see Minutes and Financial reports, Co-operative Council No. 1, Colin Fraser Papers, 4/5/6, MUA. For an outline of the welfare schemes and new facilities introduced between 1917 and 1920 see 'Port Pirie Smelters, 1920' by CDW, 22 May 1920, Colin Fraser Papers, 1/22/3/1, MUA. For the flow-on effects of these initiatives at other smelters see Eklund, 'Managers, Workers and Industrial Welfarism', pp. 150–51.
44 See 'Port Pirie Smelters, 1920' by CDW, 22 May 1920, Colin Fraser Papers, 1/22/3/1, MUA.
45 Johnstone, 'A History of the Origins', p. 328.
46 Colin Fraser, Port Pirie Zinc Smeltery Report No. 2, 4 February 1916, Colin Fraser Papers, 1/18/6/2, MUA.
47 See, for example, letter from H.St.J. Somerset to Colin Fraser, 15 September 1919, 'Strikes etc' file, BHAS Group 5 Records, 1916–1922, MUA.
48 General Superintendent's Report, Year ended 27 June 1928, part 1, p. 97 in Port Pirie – General Superintendent's Reports, BHAS Group 5 Records, MUA.
49 Letter from S.E. Fraser (Acting General Manager) to Colin Fraser, 19 February 1918, in 'Labour/Welfare' file, BHAS Group 5 Records, Port Pirie, 1915–1919, MUA.
50 See 'Welfare – attendance bonus' file, BHAS Group 5 Records, MUA.
51 For criticisms of the camp see G. Rigg to G. Mussen, 3 April 1919 in 'Welfare – Holiday Camp Re Sanatorium' file, Port Pirie Strikes etc 1916–1922, BHAS Group 5 Records, MUA. In 1926 the site and the buildings on it were sold to the Port Pirie Council. For the sale of the camp to the council see General Superintendent's Report, Year ended 27 June 1928, part 1, pp. 6–7 in Port Pirie – General Superintendent's Reports, BHAS Group 5 Records, MUA.
52 Another senior BHAS figure, Herbert Gepp – a metallurgical engineer who had worked for the Zinc Corporation, and Amalgamated Zinc, and in 1916 became General Manager of Electrolytic Zinc in Risdon – had been involved in a similar scheme at Broken Hill in 1912 and 1913 using the vehicle of a Progress Association to organise tree planting and the construction of a children's playground. Johnstone, 'A History of the Origins', pp. 313–14.
53 *Port Pirie Advertiser*, 24 February 1917.
54 *Port Pirie Advertiser*, 24 March 1917.
55 *Port Pirie Advertiser*, 2 June 1917.
56 Statement showing expenditure on Soldiers' Memorial Park and Playground period ended 25 June 1919, 'Labour Welfare – Children's Playground', 'Port Pirie, 1915–1919, Labour/Welfare', BHAS Group 5 Records, MUA.

57 M.E. Goode, *Port Pirie Mayoral Report*, 1918–1919, p. 8, Mortlock Library.
58 Robinson, *If I Remember Rightly*, p. 260.
59 See 'Welfare – BHAS Picnic and Sports Committee' file, Port Pirie 1926–28 Visitors and Welfare, BHAS Group 5 Records, MUA.
60 William Robertson to Colin Fraser, 25 January 1918, in 'Applications from Labourers etc' file, Port Pirie, 1915–1919, BHAS Group 5 Records, MUA.
61 W.S. Robinson, 'Notes on visit to Broken Hill and Port Pirie, 19 and 20 October 1915', in 'Labour' file, Port Pirie, 1915–1919, BHAS Group 5 Records, MUA.
62 W.S. Robinson to Colin Fraser, 27 July 1916 in BHAS Group 5 Records, Port Pirie, 1915–1919, MUA.
63 For the findings of the Royal Commission see Report of the Royal Commission on Plumbism Together with Minutes of Evidence and Appendices, *South Australian Parliamentary Papers*, vol. 2, 1925. For the focus on foreign workers see R. Gillespie, 'Accounting for lead poisoning: The medical politics of occupational health', *Social History*, vol. 15, no. 3, 1990, pp. 320–21. For the figures on foreign workers see General Superintendent's Report for the Year ended 27 June 1928, part 1, p. 97 and part 2, p. 15 in 'Port Pirie General Superintendent's Reports', BHAS Group 5 Records, MUA.
64 See Johnstone, 'A History of the Origins', p. 215. See also *Australian Worker*, 17 April 1917, p. 16.
65 William Robertson to Colin Fraser, 25 May 1918, Port Pirie, 1915–1919, BHAS Group 5 Records, MUA.
66 'Port Pirie Smelters, 1920', p. 21.
67 John Wanna, *Defence not Defiance: The development of organised labour in South Australia*, Texts in Humanities, Adelaide, 1981, pp. 61–62.
68 In 1920 the state secretary of the AWU, Frank Lundie, argued that the AWU stood for 'political action, arbitration and conferences for settling disputes'. See NBAC/ANU, AWU, Adelaide and South Australian Branch, E154/58, Secretary's Report for the Year Ending 30 May 1920, p. 8, Annual Reports and Balance Sheets 1895–1957.
69 *Port Pirie Recorder*, 9 August 1917.
70 Johnstone, 'A History of the Origins', p. 324.
71 *Port Pirie Recorder*, 28 June 1917, editorial.
72 *The Recorder*, 6 January 1923.
73 *Port Pirie Recorder*, 22 January 1921.
74 W. Robinette, AWU Handbill in 'Port Pirie, 1922–1926', BHAS Group 5 Records, MUA. See also Gillespie, 'Accounting for lead poisoning', pp.303–22. For Robinette's opposition see also Gillespie, 'Accounting for lead poisoning', pp. 313–14, 320–23.
75 'Welfare – Dental Clinic', 'Port Pirie, 1926–1928', BHAS Group 5 Records, MUA.
76 Gerald Mussen to Manager, BHAS, Melbourne, 16 May 1924, 'Welfare', '1922–1925', BHAS Group 5 Records, MUA.
77 Gerald Mussen to Colin Fraser, 23 July 1918, in 'Labour Welfare – Children's Playground' file, Port Pirie, 1915–1919, Labour/Welfare, BHAS Group 5 Records, MUA.
78 Johnstone, 'A History of the Origins', p. 340.
79 'Welfare' file, 1922–1925, BHAS Group 5 Records, MUA.
80 *The Recorder*, 17 January 1923.

81 'Port Pirie Smelters, 1920', pp. 1–10.
82 Letter from W.A. Woodward to Secretary, BHAS, Melbourne, 4 December 1925, in 'BHAS Co-op Council Constitution' file, 'Port Pirie, 1925–26', BHAS Group 5 Records, MUA.
83 NBAC/ANU, Amalgamated Metal Workers Union, South Australian Branch, E200/7, Minutes of the Port Pirie Branch of the AEU, 9 October 1924.
84 Letter from W.A. Woodward to Secretary, BHAS, Melbourne, 12 August 1936, in 'Company papers: welfare Co-op store to 1936', BHAS Group 5 Records, MUA.
85 This argument is developed by M.L. Wingerd, 'Rethinking Paternalism: Power and paternalism in a southern mill village', *Journal of American History*, vol. 83, 1992, pp. 872–902. In the southern US textile town of Cooleemee, Wingerd argues that 'below the surface of apparently placid labor relations an elaborate culture of resistance operated within a locally circumscribed system of negotiation': p. 874.
86 J.F. Jenkins, *Port Pirie Mayoral Report, 1920–1921*, p. 8, Mortlock Library, South Australia.
87 J.F. Jenkins, *Port Pirie Mayoral Report, 1921–1922*, p. 9, Mortlock Library, South Australia.
88 *Port Pirie Advertiser*, 20 October 1917.
89 See advertisement for the 1932 Port Pirie Waterside Workers Federation Picnic reproduced in A. Reeves, *Another Day Another Dollar: Working class lives in Australian history*, McCullock Publishing, Melbourne, 1988, p. 91.
90 NBAC/ANU, Waterside Workers Federation Branch Correspondence, T62/36/2, Branch Balance Sheets Sent to the Federal Office, Port Pirie Waterside Workers Federation, Accounts for the Year Ended 31 December 1928.
91 *Port Pirie Advertiser*, 6 October 1917.
92 The date of the opening of the store is uncertain. The branch approved the idea by 188 votes to 26 in July 1920, and the store was located in the AWA Hall in Florence Street. The store continued operating until 1956. See NBAC/ANU, Waterside Workers Federation Branch Correspondence, T62/12/4/2; and *Port Pirie Conservation Study*, Lester Firth and Murton Pty Ltd, Adelaide, September 1980.
93 Ken Bullock, *Port Pirie: The friendly city – The undaunted years*, Peacock Publications, Norwood (SA), 1988, p. 114.
94 Desmond O'Connor, *No Need To Be Afraid: Italian settlers in South Australia between 1939 and the Second World War*, Wakefield Press, Adelaide, 1996, p. 69.
95 Donley, *The Rise of Port Pirie*, pp. 18–22.
96 See also Bullock, *Port Pirie: The friendly city*, p. 293.
97 O'Connor, *No Need To Be Afraid*, pp. 68–79; and Graeme Hugo, 'Patterns and Processes of Italian Settlement in South Australia', in Des O'Connor and Antonio Comin (eds), *The First Conference on the Impact of Italians in South Australia (16–17 July 1993): Proceedings*, Italian Congress Inc. and Flinders University, Adelaide, 1993, pp. 33–66.
98 O'Connor, *No Need To Be Afraid*, pp. 76–77.
99 Ibid., pp. 76–77.
100 See Gillespie, 'Accounting for Lead Poisoning', p. 322.
101 Interview with Bruce Steel, p. 3.
102 Des O'Connor, 'The post-war settlement of Italians in South Australia', in Des

O'Connor (ed.), *Memories and Identities: Proceedings of the Second Conference on the Impact of Italians in South Australia*, Australian Humanities Press, Adelaide, 2004, pp. 70–71.
103 *Census of the Commonwealth of Australia, 30 June 1954*, Vol. IV, Part 1 'Analysis of Population in Local Government Areas', pp. 24–25.
104 *The Advertiser* (Adelaide), 2 March 1953, p. 2.
105 *The Advertiser* (Adelaide), 3 April 1926, p. 18.
106 R.G. Sandercock, 'Thirty Years of Lead Refining at the BHAS Pty Ltd, Port Pirie, South Australia', *Supplement to the Australian Institute of Mining and Metallurgy*, October, 1951.
107 *Census of the Commonwealth of Australia, 30 June 1966*, Vol. 4, Part 4 'Population and Dwellings in Local Government Areas', pp. 64–65.
108 *Census of the Commonwealth of Australia, 30 June 1954*, 'No. 7. – Occupational Status: Local government areas and non-municipal towns (1,000 persons or more), South Australia, 30 June 1954', p. 88.
109 *Port Pirie Conservation Study*, 2.2 and 2.3.
110 Bullock, *Port Pirie: The friendly city*, pp. 52–54.
111 *The Advertiser* (Adelaide), 23 January 1953, p. 1. I would like to thank local historian Bob Krahnert for sharing this information and for showing me his extensive collection of Port Pirie photographs and memorabilia.
112 Frank Green, *The Port Pirie Smelters*, Broken Hill Associated Smelters, Melbourne, 1977, p. 110.
113 Interview with Bruce Steel, pp. 5–6.
114 Biographical details compiled by R.J. Hopkins and published in the preface to Green's history of BHAS; Green, *The Port Pirie Smelters*.
115 *Barrier Miner*, 25 April 1947, p. 8.
116 Bullock, *Port Pirie: The friendly city*, p. 341; and Geoffrey Blainey: *Steel Master: The life of Essington Lewis*, Macmillan, Melbourne, 1971, p. 71.
117 Blainey, *Steel Master*, pp. 31–45.
118 J.S. Geddes served as mayor of Port Pirie in 1915–16, 1919–20 and 1924–25. He was the son of Charles Geddes, former alderman and mayor, local auctioneer, and one of the earliest residents of Port Pirie.
119 Jayne Mayo Carolan, *No Run-of-the-Mill: A biography of Henry Beaufort Somerset*, Australian Scholarly Publishing, Melbourne, 2006. Henry ('Harry') Beaufort Somerset was born in 1906 and died in 1995.
120 Interview with Leonard Zubrinich (by Edward Stokes in the Broken Hill social history project. 26 June 1982), NLA.
121 Joy Nunan, cited in Bullock, *Port Pirie: The friendly city*, p. 56.
122 *The Mail* (Adelaide), 11 September 1937, p. 2.
123 For a survey of Whyalla's history and economic development see 'Regional Study – Whyalla', *South Australian Year Book*, Australian Bureau of Statistics – South Australian Office, Adelaide, No. 29, 1995, pp. 332–51.
124 Bullock, *Port Pirie: The friendly city*, pp. 27, 109.
125 R. Keith Johns, 'Uranium in South Australia – Politics and Reality', *Journal of Australasian Mining History*, vol. 3, September 2005, pp. 171–84.
126 See M. McLeary, *Port Pirie Uranium Treatment Plant: Management Plan, Phase 1 – Preliminary Investigation*, Primary Industries and Resources, South Australian Government, Adelaide, 2004. Various other activities were carried out on-site after 1962, including a rare earth treatment plant, but these have been excluded

here for reasons of space.
127 *Commonwealth Year Book*, 1975, p. 1110.
128 Donley, *The Rise of Port Pirie*, p. 7.
129 Green, *The Port Pirie Smelters*, p. 159.
130 B. Cartwright, R.H. Merry and K.G. Tiller, 'Heavy metal contamination of soils around a lead smelter at Port Pirie, South Australia', *Australian Journal of Soil Research*, vol. 15, no. 1, 1976, pp. 69–81.
131 Ibid., p. 80.
132 E.J. Maynard, L.J. Franks and M.S. Malcolm, *The Port Pirie Lead Implementation Program, Future Focus and Directions*, Department of Health, Adelaide, 2005, p. 1.
133 Interview with Jenny Pryor.

6 Mount Isa: Normalising outback suburbia

1 Ray Evans, *A History of Queensland*, CUP, Cambridge, 2007, p. 216. While drafting this chapter I was fortunate enough to read a manuscript version of Neil White's historical comparison of Mount Isa and Corner Brook, Canada, subsequently published as *Company Towns: Corporate order and community*, University of Toronto Press, Toronto, 2011. White's analysis is conceptually different and presents an argument which departs from this chapter at key points. However, the book is an important addition to the historiography of Mount Isa, particularly with the inclusion of an international comparative dimension. Another late arrival was Noreen Kirkman's PhD thesis, ('Mount Isa Mines' Social Infrastructure Programs, 1924–1963', James Cook University, 2011), which is another valuable addition to the historiography of Mount Isa.
2 Australian Bureau of Meteorology, 'Climate Statistics for Australian Locations', Mount Isa Mine site (1932–2011) www.bom.gov.au/climate/data/.
3 Kirkman, *Mount Isa: Oasis of the outback*, pp. 4–5.
4 D. Trigger, *Whitefella Comin': Aboriginal responses to colonialism in northern Australia*, CUP, Cambridge, 1992, pp. 17–37.
5 *Townsville Daily Bulletin*, 23 November 1935 p. 7.
6 Don Berkman, *Making the Mount Isa Mine: 1923–1933*, The Australasian Institute of Mining and Metallurgy, Spectrum Series no. 1, Melbourne, 1996. p. 126; and B. Penrose, 'Occupational Lead Poisoning at Mount Isa Mines in the 1930s', *Labour History*, no. 73, November 1997, p. 126.
7 For information on Camooweal and its retail history see Queensland State Heritage Listing for Freckleton's Stores, Camooweal, www.epa.qld.gov.au/projects/heritage/index.cgi?place=600739&back=1.
8 Raphael Cilento and Clem Lack, *Triumph in the Tropics: An historical sketch of Queensland*, The Historical Committee of The Centenary Celebrations Council of Queensland, Brisbane, 1959, pp. 385–86.
9 Kirkman, *Mount Isa: Oasis of the Outback*, p. 6.
10 *Townsville Daily Bulletin*, 8 April 1929, p. 6; and *Brisbane Courier*, 3 December 1929, p. 10.
11 *Sydney Morning Herald*, 18 June 1924, p. 14.
12 Kirkman, *Mount Isa: Oasis of the outback*, p. 7.
13 Ibid., p. 8; and *Cairns Post*, 8 May 1924, p. 7.
14 The Mount Isa Chamber of Commerce was later reformed and was active in the 1930s and 1940s. It played a very similar role to the Progress Association, and in 1935 the Chamber actually absorbed the Progress Association. See *Townsville*

Daily Bulletin, 8 February 1935, p. 4.
15 Interview with M. Saslavski (by Noreen Kirkman in 1984), cited in Kirkman, *Oasis of the outback*, p. 19.
16 *Sydney Morning Herald*, 3 May 1924, p. 18.
17 Penrose, 'Occupational Lead Poisoning at Mount Isa Mines', pp. 123–41.
18 Berkman, *Making the Mount Isa Mine*, pp. 8–10.
19 *35 Years of the Q.C.W.A in Mount Isa, 1928–1963*, Mount Isa, Queensland CWA Branch, 1963; and E.K. Teather, 'Remote Rural Women's Ideologies, Spaces and Networks: Country Women's Association of New South Wales, 1922–1992', *Journal of Sociology*, vol. 28, no. 3, 1992, pp. 369–90.
20 *Townsville Daily Bulletin*, 26 August 1929, p. 9.
21 Information for this section has been sourced from the following newspaper reports: *Townsville Daily Bulletin*, 27 April 1929, p. 7; *Courier Mail*, 12 April 1932, p. 4; *Courier Mail*, 28 August 1948, p. 5.
22 Interview with E.T. 'Taffy' Badham (by Carol Gistitin, 5 August 1989) Capricornia Collection, CQU, p. 5.
23 Blainey, *Mines in the Spinifex*, p. 157.
24 See, for example, Barry McGowan, *Dust and Dreams: Mining communities in south-east New South Wales*, UNSW Press, Sydney, 2010, p. 83.
25 For a small sample of the innumerable reports about Mount Isa in the Broken Hill press see *Barrier Miner*, 2 May 1924 (mining developments), p. 1; *Barrier Miner*, 4 February 1933 (recent strike action) p. 3; and *Barrier Miner*, 5 October, 1954 (appointment of J.W. Foot as mine manager), p. 2.
26 Penrose, 'Occupational Lead Poisoning at Mount Isa Mines', p. 125.
27 *Mount Isa Mine Ltd*, Mount Isa, August 1939, MIM Ltd.
28 Dianne Menghetti, 'Mount Isa: A Town Like Alice?', *Australian Historical Studies*, vol. 109, October 1997, pp. 21–32. See also Penrose, 'Occupational Lead Poisoning at Mount Isa Mines', p. 125.
29 Berkman, *Making the Mount Isa Mine*, pp. 129–37.
30 *Mount Isa Mine Ltd*, Mount Isa, August 1939, MIM Ltd.
31 J. Wagner, 'Why garden?: Gardening on mining fields in the dry tropics of Queensland, 1860 to 1960', *Journal of Australian Studies*, vol. 34, no. 3, 2010, p. 356.
32 Donald Chaput and Kett Kennedy, *The Man from ASARCO: A life and times of Julius Kruttschnitt*, Australasian Mineral Heritage Trust/Australasian Institute of Mining and Metallurgy, Melbourne, 1992, pp. 152–54; and *Townsville Daily Bulletin*, 13 April 1929, p. 4.
33 Cited in Chaput and Kennedy, *The Man from ASARCO*, p. 136.
34 *San Fransicso Call*, 25 September 1907, p. 1.
35 *Townsville Daily Bulletin*, 6 November 1931, p. 6 and *Courier Mail*, 5 September 1940, p. 9.
36 *Quin's Metal Handbook and Statistics*, London, Metal Information Bureau, 1953; F.E. Banks, *Copper in the World Economy*, Ballinger Publishing Company, Cambridge MA, 1974, p. 12; and *Official Year Book of the Commonwealth of Australia*, no. 44, 1955, p. 764.
37 *Courier Mail*, 15 November 1952, p. 6.
38 Douglas Blackmur, 'Arbitration, Legislation and Industrial Peace: Queensland in the Reconstruction Years', *Labour History*, no. 63, November 1992, pp. 115–34.
39 Mount Isa Mines Ltd, *Annual Report for Year 1970*, p. 18.

40 Cited in Chaput and Kennedy, *The Man from ASARCO*, p. 136.
41 Interview with Margaret Jones, p. 1.
42 Interview with Sigge Lampen, p. 6.
43 Interview with Margaret Jones, p. 8.
44 See, for example, Robert Dixon and Veronica Kelly (eds), *Impact of the Modern: Vernacular modernities in Australia, 1870s–1960s*, Sydney University Press, Sydney, 2008. This is an excellent collection, but the absence of any consideration of capitalism as both a driving force and central experience of modernity is an oversight. Did 'ordinary' people not experience, resist, or progress modernity?
45 Interview with Betty Addison.
46 Ibid.
47 Interview with Kalle Nurmi.
48 Interview with Sandy Jones.
49 Interview with Noel Addison.
50 *Official Year Book of the Commonwealth of Australia*, no. 55, 1969, p. 1037.
51 Blainey, *Mines in the Spinifex*, pp. 222–23.
52 *MIMAG*, no. 37, October 2006, p. 10.
53 Pat Mackie (with Elizabeth Vassilieff), *Mount Isa: The story of a dispute*, Hudson, Hawthorne (Vic), 1989, pp. 3–4.
54 Blainey, *Mines in the Spinifex*, pp. 214–15.
55 *Barrier Daily Truth*, 8 March 1951, p. 2; *Morning Bulletin*, 6 February 1951, p. 1; and Table 5.10 'Average weekly earnings (a), male employees by State and Territory, Australia, 1945–46 to 1996', in *Queensland: Past and present: 100 Years of Statistics, 1896–1996*, Office of Economic and Statistical Research, Queensland Government, Brisbane, 1996, p. 141.
56 *Courier Mail*, 26 July 1951, p. 2.
57 Cited in Bretta Carthey and Bob Potter, *Mount Isa: the Great Queensland strike*, Solidarity Pamphlet No. 22. libcom.org/library/mount-isa-great-queensland-strike-solidarity.
58 Mackie, *Mount Isa: The story of a dispute*, pp. 1–27.
59 *Census of the Commonwealth of Australia, 30 June 1954*, Vol. III, Queensland, Part 1 'Analysis of Population in Local Government Areas', pp. 149, 163.
60 Interview with Sandy Jones.
61 Evidence of the MIM Chief Auditor, C.G. Firmin to the Tariff Board Case, cited in *Cairns Post*, 27 March 1953, p. 1.
62 *Morning Bulletin*, 23 July 1953, p. 1 and Queensland Places website – entry for 'Mount Isa suburbs', queenslandplaces.com.au/home.
63 Interview with Margaret Jones, p. 10.
64 See entry in the Queensland Heritage Register, Place ID 601094, www.epa.qld.gov.au/chims/basicSearch.html.
65 *Townsville Daily Bulletin*, 16 July 1932, p. 16.
66 See entry in the Queensland Heritage Register, Place ID 600742, www.epa.qld.gov.au/chims/basicSearch.html.
67 P. Memmott, 'From the 'Curry to the 'Weal: Aboriginal town camps and compounds of the western back-blocks', *Fabrications*, no. 7, August 1996, pp. 20–21.
68 M. Moran and D. Burgen, *Gulf and West Queensland Regional Homeland Plan: Stage II Main Report*, Gulf Aboriginal Development Company/ATSIC Gulf and West Queensland Regional Council, Mount Isa, 2000, pp. 16–17.

69 Mount Isa Welfare Council, *A Report into Some Aspects of Deprivation in an Area of North-West Queensland*, Mount Isa Welfare Council, Mount Isa, 1976, p. 2.
70 J. Hagan and R. Castle, 'Regulation of Aboriginal labour in Queensland: Protectors, Agreements and Trust Accounts, 1897–1965', *Labour History*, no. 72, May 1997, pp. 66–76.
71 Moran and Burgen, *Gulf and West Queensland Regional Homeland Plan*, p. 13.
72 *Courier Mail*, 6 December 1939, p. 2.
73 Interview with Kalle Nurmi.
74 *Census of the Commonwealth of Australia, 30 June 1966*, Vol. IV, Queensland, Part 1 'Population and Dwellings in Local Government Areas', pp. 104–05. The large group of Finns was not classified separately and fell into the 'Other European' category.
75 R. Kirkpatrick, 'How a communist rag in Darwin became a newspaper for Mount Isa', *PANPA Bulletin*, June 2005, pp. 56–57.
76 *Mount Isa Today*, MIM Ltd, Mount Isa, 1967.
77 Ibid. See also Kirkman, *Mount Isa: Oasis of the outback*, pp. 93–105.
78 *Census of the Commonwealth of Australia, 30 June 1954*, Vol. III, Queensland, Part 1 'Analysis of Population in Local Government Areas', p. 163.
79 Mackie, *Mount Isa: The story of a dispute*, p. 17.
80 Cited in Menghetti, 'A Town Like Alice?', p. 24.
81 M. Cribb, 'The Mount Isa strike, 1961, 1964', in D.J. Murphy (ed.), *The Big Strikes: Queensland, 1889–1965*, UQP, Brisbane, 1983. See also G.H. Sorrell, 'The Dispute at Mount Isa', *Australian Quarterly*, no. 36, 1965, pp. 22–33.
82 Mackie, *Mount Isa: The story of a dispute*, pp. 64–65.
83 Interview with Sandy Jones.
84 Interview with Lawrence Miller (by Alex Hood, 29 May 2002), Oral TRC 4864/4-5, NLA.
85 Mackie, *Mount Isa: The story of a dispute*, p. 6.
86 Ibid., p. 19.
87 Ibid., p. 53; and Hearn and Knowles, *One Big Union: A history of the Australian Workers' Union, 1889–1994*, pp. 258–77.
88 Interview with Sir Frank Nicklin (by Suzanne Lunney, 19 March 1974), Oral TRC 254. NLA. Nicklin was Premier of Queensland from 1957 to 1968. The company's view of the dispute is found in Gordon Sheldon, *Industrial Siege*, Cheshire, Melbourne, 1965. Sheldon was a former public relations consultant for MIM, a fact that remains undeclared in the book.
89 Mackie, *Mount Isa: The story of a dispute*, p. 98.
90 Interview with Sandy Jones.
91 *Mount Isa Today*, MIM Ltd, Mount Isa, 1967.
92 Doug Hunt, 'Lockout at Mount Isa', in M. Gardner and G. Palmer (eds), *Employment Relations: Industrial relations and human resource management in Australia*, Macmillan, Melbourne, 1992, pp. 508–19.
93 Interview with Margaret Jones, p. 3.

7 Kambalda: Modernity, environment and experience

1 Michelle Handley, 'The distribution pattern of algal flora in Saline lakes in Kambalda and Esperance, Western Australia', MSc thesis, Curtin University, 2003, p. 29. The 242mm figure for annual rainfall at Kambalda is from J.D.A. Clarke, 'Ancient Landforms of Kambalda and their significance to human

activity', in V.A. Gostin (ed.), *Gondwana to Greenhouse: Australian environmental geoscience*, Geological Society of Australia Inc., Special Publication No. 21, Southwood Press, Sydney, 2001, p. 50.
2. Cited in J.J. Gresham, *Kambalda: History of a mining town*, Western Mining Corporation Limited, Melbourne, 1991, p. 4. For early European exploration generally see ibid., pp. 1–13.
3. Charles Hunt, cited in Gresham, *Kambalda: History of a mining town*, p. 10.
4. Cited in Gresham, *Kambalda: History of a mining town*, p. 25.
5. Cited in Norma King, no page numbers (1). See also Norma King, *Nickel Country – Gold Country*, Rigby, Adelaide, 1972, pp. 7–8.
6. *Leader* (Melbourne newspaper), cited in Gresham, *Kambalda History*, pp. 28–29.
7. Lindesay Clark, *Built on Gold: Recollections of Western Mining*, Hill of Content, Melbourne, 1983, p. 202.
8. Australian Bureau of Statistics, 'Australian Historical Population Statistics, Table 30. Population (a), age and sex, WA, 1901 onwards', www.abs.gov.au/AUSSTATS/abs@.nsf/DetailsPage/3105.0.65.00120042OpenDocument.
9. Clark, *Built on Gold*, p. 201; and Gresham, *Kambalda History*, p. 54.
10. Western Mining Corporation formally changed its name to WMC Resources Ltd in 1996; for the sake of brevity the acronym WMC will be used throughout this book to cover all periods of the company's history.
11. Clark, *Built on Gold*, p. 202; and Trevor Sykes, *The Money Miners: The great Australian boom, 1969–1970*, Wildcat Press, Sydney, 1978, pp. 7–10.
12. T. Stannage (ed.), *A New History of Western Australia*, UWA Press, Perth, 1981, p. 461.
13. See, especially, Patrick Bertola, 'Kalgoorlie, Gold, and the World Economy 1893–1972', PhD thesis, Curtin University of Technology, WA, 1993; and Naomi Segal, 'Compulsory Arbitration and the Western Australian Gold-Mining Industry: A re-Examination of the inception of compulsory arbitration in Western Australia', *International Review of Social History*, vol. 47, no. 1, 2002, pp. 67–69.
14. 'Nickel Statistics, US Geological Survey', 16 December 2008.
15. L.C. Brodie-Hall and J.B. Oliver, 'Kambalda Case Studies-1', p. 188, in Australian UNESCO Seminar, *Man and the Environment: New towns in isolated settings*, Australian Government Publishing Service, Canberra, 1976; Dean M. Hoatson, Subhash Jaireth and A. Lynton Jaques, 'Nickel sulfide deposits in Australia: Characteristics, resources, and potential', *Ore Geology Reviews*, vol. 29, Issues 3–4, November 2006, p. 179.
16. *Australia's Identified Mineral Resources, 2008*, Geoscience Australia, Australian Government, Canberra, 2008, pp. 51–52.
17. *Sunday Times*, 15 May 1966.
18. Jack Lunnon, cited in King, *Nickel Country – Gold Country*, p. 13.
19. Brodie-Hall and Oliver, 'Kambalda Case Studies-1', p. 188.
20. See Meredith Fletcher, *Digging Up People for Coal: A history of Yallourn*, MUP, Melbourne, 2002.
21. Brodie-Hall and Oliver, 'Kambalda Case Studies-1', p. 193.
22. J. McCulloch, 'The Mine at Wittenoom: Blue asbestos, labour and occupational disease', *Labor History*, vol. 47, no. 1, 2006, pp. 1–2.
23. *West Australian*, 4 August 1966.
24. Elspeth Huxley, *Their Shining Eldorado: A journey through Australia*, Chatto &

Windus, London, 1967, p. 190.
25 King, *Nickel Country – Gold Country*, pp. 1–2.
26 Ibid., p.7.
27 Interview with Will Manos.
28 *Nickel Refinery (Western Mining Corporation Limited) Agreement Act 1968* (WA).
29 See, for example, *Kanowna Heritage Trail*, Heritage Council of Western Australia/ Delta, Gold N.L., 1993. Kanowna was formerly known as 'White Feather'.
30 For a general discussion of the way in which mining development is viewed see David Trigger, 'Mining, landscape and the culture of development ideology in Australia', *Ecumene*, vol. 4, no. 2, 1997, pp. 167–69.
31 *WAPD*, vol. 175, 1967, p. 120 and vol. 180, 11 September 1968.
32 Clark, *Built on Gold*, p. 206; and G.M. Ralph, *A Pictorial History of Kambalda*, WMC, Melbourne, 1992, p .16.
33 *Nickel Refinery (Western Mining Corporation Limited) Agreement Act 1968* (WA).
34 Letter from K.N to Peter Lucas, 3 October 1969, Engineering Services, Town Planning 11/1/1968–31/12/1969 No. 1, Box 230, WMC Collection, UMA.
35 Ibid.
36 J.B. Oliver, 'Aspects of Town Development at Kambalda', Engineering Services, Town Planning 11/1/1968–31/12/1969 No. 1, Box 23, WMC Collection, UMA.
37 G.M. Ralph, *A Pictorial History of Kambalda*, Western Mining Corporation Limited, Melbourne, 1992, pp. 16–20.
38 See especially G.C. Borrack, 'Preliminary Report: Proposed research into new building and township development in Western Australia for Western Mining Corporation Limited', 1969?, Engineering Services, Town Planning 11/1/1968–31/12/1969 No. 1, Box 230, WMC Collection, UMA. Borrack was WMC's in-house architect; he had joined the company in 1969.
39 See C. Lee and B. Stabin-Nesmith, 'The Continuing Value of a Planned Community: Radburn in the evolution of suburban development', *Journal of Urban Design*, vol. 6, no. 2, 2001, pp. 152–54.
40 Philip Goad, 'Mary Kathleen and Weipa: Two model townships for post-war Australia', *Transition*, vols 49/50, 1996, pp. 49–52.
41 Interview with David Stokes and interview with Will Manos.
42 Gresham, *Kambalda: History of a mining town*, pp. 82, 86.
43 Interview with David Stokes.
44 Ibid.
45 'Nickel Outshines Gold', *Australian Women's Weekly*, 5 December 1973, p. 10. See also 'John Jones Finds a Fortune', *Australian Women's Weekly*, 25 September 1968, p. 10.
46 John Simon, 'Three Australian Asset-price Bubbles', in Anthony Richards and Tim Robinson (eds), *Proceedings of the Asset Prices and Monetary Policy – Reserve Bank Annual Conference*, 18–19 August 2003, Sydney, pp. 22–30, www.rba.gov.au/ publications/confs/2003/index.html.
47 Simon, 'Three Australian Asset-price Bubbles', p. 29.
48 O.A. Oeser, 'Preface', in Australian UNESCO Seminar, *Man and the Environment, New towns in isolated settings*, p. ix.
49 See, for example, O.A. Oeser, 'Single Industry Company Towns: The psychology of community development', in M. Bowman (ed.), *Beyond the City: Case studies in community structure and development*, Longman Cheshire, Melbourne, 1981, pp. 203–17.

Notes to pages 221–238

50. P.J.R. Redding had a BA (Hons) in Psychology and Philosophy and a Diploma in Social Studies; J.K. Austin had a BA (Hons) in Psychology. See Australian UNESCO Seminar, *Man and the Environment: New towns in isolated settings*, pp. 201, 215.
51. Mark Healy and Lyne Acacio, *Regional Social Indicators for Aboriginal People in Western Australia*, Department of Aboriginal Affairs, Perth, 1996, p. 24.
52. P.J.R. Redding, 'Kambalda Case Studies-3 Living in Kambalda', p. 203, in Australian UNESCO Seminar, *Man and the Environment: New towns in isolated settings*.
53. Interview with Will Manos and interview with David Stokes.
54. Interview with Norma Worthington, pp. 3–5.
55. Interview with David Stokes.
56. Redding, 'Kambalda Case Studies-3 Living in Kambalda', p. 204.
57. Interview with Norma Worthington, p. 5.
58. J. Skivinis and A. Taylor, *A Mining Town: Kambalda*, Longman, Melbourne, 1975, p. 12.
59. See generally, *Western Mining Corporation, Annual Reports*, 1968–1974, State Library of Victoria.
60. Interview with David Stokes.
61. Interview with Will Manos.
62. *Western Mining Corporation*, Annual Report for the Year 1971, p. 15ff.
63. Interview with Norma Worthington.
64. Gresham, *Kambalda: History of a mining town*, pp. 93–97.
65. *WAPD*, vol. 218, 1977, p. 1208.
66. *WAPD*, vol. 239, 1982, pp. 2295–96.
67. *Sydney Morning Herald*, 6 July 1991.
68. *Green Left Weekly*, 27 November 1991.
69. *WAPD*, vol. 358, 1999–2000, p. 58.
70. *1996 Commonwealth Census*, 2016.5 – Census of Population and Housing: Selected Characteristics for Urban Centres and Localities, Western Australia, Cocos (Keeling) and Christmas Islands, 1996, www.abs.gov.au.
71. *2001 Commonwealth Census*, 2016.0 – Census of Population and Housing: Selected Characteristics for Urban Centres, Australia, 2001, www.abs.gov.au.
72. See also a critical analysis of the rehabilitation project carried out at the Mary Kathleen mine in Queensland, another planned town. B.G. Lottermoser, P.M. Ashley and M.T. Costelloe, 'Contaminant dispersion at the rehabilitated Mary Kathleen uranium mine, Australia', *Environmental Geology*, vol. 48, no. 6, 2005, pp. 748–61.

Postscript

1. See David Harvey, 'Spatial Fix and Globalization', *Geographische Revue*, vol. 2, 2001, pp. 23–30.
2. The fate of these small, often ephemeral, mining communities is best evoked by McGowan *Dust and Dreams*.
3. See, for example, *Broken Hill, NSW, Australia: Heritage & cultural tourism initiatives, 1986–2005, Information for the Productivity Commission Enquiry*, July 2005, www.pc.gov.au/__data/assets/pdf_file/0018/54162/sub012.pdf.

Appendix

1. The relevant approval numbers are Newcastle (H-009-0405) and Monash (CF08/0944 – 2008000459).
2. I have explored these and other issues, including the definition of vernacular history, in a number of journal articles with various co-authors. See A. Eklund and E. Eklund, '"Do you Love the Town you Live in?": Narratives of place from Australian mining towns', *The International Journal of Interdisciplinary Social Sciences*, vol. 3, no. 7, 2008, pp. 53–57; and E. Caprioni, M. Deiana and E. Eklund, 'Interview Techniques in Three Different Research Scenarios', *The International Journal of Interdisciplinary Social Sciences*, vol. 4, no. 1, 2009, pp. 1–10.

Bibliography

Primary sources

Commonwealth and state government publications

Colonial Census Reports from New South Wales, Queensland, South Australia, Tasmania and Western Australia, *ANU Historical Census and Data Archive*, hccda.anu.edu.au/documents/.

Census of the Commonwealth of Australia, 1901, 1911, 1921, 1933, 1947, 1955, 1966, 1976, 1981, 1991, 1996, 2001, 2006, www.abs.gov.au.

Coghlan, Timothy, *General Report on the Eleventh Census of New South Wales*, D. Potter, Government Printer, Sydney, 1894.

Commonwealth Year Book, No. 1, 1908.

Official Year Book of the Commonwealth of Australia, No. 40, 1954.

Official Year Book of the Commonwealth of Australia, No. 44, 1955.

Official Year Book of the Commonwealth of Australia, No. 46, 1960.

Official Year Book of the Commonwealth of Australia, No. 55, 1969.

Official Year Book of the Commonwealth of Australia, No. 56, 1970.

Official Year Book of the Commonwealth of Australia, No. 61, 1975.

Report of the Royal Commission on Plumbism Together with Minutes of Evidence and Appendices, *South Australian Parliamentary Papers*, Vol. 2, 1925.

South Australian Year Book, Australian Bureau of Statistics – South Australian Office, Adelaide, No. 29, 1995.

Company records

Broken Hill Associated Smelters Pty Ltd Records, Group 5 Records, 1928–1940, Melbourne University Archives.

Broken Hill Proprietary Ltd, *Annual Reports*, 1918–1936, Mitchell Library, Sydney.

Broken Hill Proprietary Ltd Records, 1927–1940, BHP Archives (this facility has since been disbanded).

Mount Isa Mines Ltd, *Annual Reports*, 1945–1985, Victorian State Library.

Mount Lyell Mining and Railway Company Ltd, *Half Yearly Reports and Balance Sheets*, 1898–1915, Victorian State Library.

Mount Lyell Mining and Railway Company Records, NS 1711/934/NS1911/91, Tasmanian State Archives.

Mount Morgan Manuscript Collection, MS D14/1281.1, Capricornia Collection, Central Queensland University (CQU).

Mount Morgan Gold Mine Company Ltd, *Annual Report*, 1886, Mitchell Library, Sydney.

Mount Morgan Gold Mine Company Ltd, *Annual Reports*, 1899–1906, Mount

Morgan Manuscript Collection, MS D15/271.1, Capricornia Collection, Central Queensland University.

Mount Morgan Gold Mining Company Limited, *Directors' Report to Shareholders*, 1888, 1925, 1935, Mitchell Library, Sydney.

Mt Morgan Gold Mining Company, *Reports of the Ordinary Annual Meeting*, 1902 and 1905–1912, Victorian State Library.

Sir Colin Fraser Papers, Melbourne University Archives.

W.S. Robinson Papers, Melbourne University Archives.

Western Mining Corporation Pty Ltd, *Annual Reports*, 1960 to 1996, Victorian State Library.

Western Mining Corporation Collection, Box 230, Melbourne University Archives.

Union records

Australian Workers' Union, Adelaide and South Australia Branch, E154/58, State Secretary's Report for the year Ended 31 May 1917, Noel Butlin Archives, ANU.

Amalgamated Engineering Union, E200/7, Minutes of the Port Pirie Branch of the ASE, 2 July 1915–26 November 1923, Noel Butlin Archives, ANU.

Waterside Workers' Federation Branch Correspondence, T62/12/4/2, Waterside Workers Federation Branch Correspondence, Noel Butlin Archives, ANU.

Amalgamated Engineering Union, Minutes of the Federated Society of Iron Shipbuilders and Boilermakers, 28 August 1918–16 May 1928, Noel Butlin Archives, ANU.

AWU, Adelaide and South Australian Branch, E154/58, Noel Butlin Archives, ANU.

Amalgamated Metal Workers Union, South Australian Branch, E200/7, Minutes of the Port Pirie Branch of the AEU, 9 October 1924, Noel Butlin Archives, ANU.

Waterside Workers Federation Branch Correspondence, T62/36/2, Branch Balance Sheets Sent to the Federal Office, Port Pirie Waterside Workers' Federation, Accounts for the Year Ended 31 December 1928, Noel Butlin Archives, ANU.

Oral history – Interviews from other sources

Interview with Emmanuel Attard (by Barry York, 15 March 1982), Oral TRC 2406/1, National Library of Australia (NLA).

Interview with E.T. 'Taffy' Badham (by Carol Gistitin, 5 August 1989), Capricornia Collection, CQU.

Transcript of interview with Arthur Barnham (by B. Cosgrove, 23 April 1993), Capricornia Collection, CQU.

Transcript of interview with George Beattie (by Lorna McDonald, 20 November 1992), Capricornia Collection, CQU.

Interview with Doris Bell (by Ed Stokes, 10 January 1982), Oral TRC 1873/6–7, NLA.

Interview with Bill Butcher (by Betty Cosgrove, 16 October 1992), Capricornia Collection, CQU.

Interview with Charles Copeman (by John Farquharson, 8 and 10 February 2000), Oral TRC 3976, NLA.

Interview with Clem Davison (by Edward Stokes in the Broken Hill social history project, 8–9 August 1982), Oral TRC 1873/94–95, NLA.

Interview with Bryce Hargreaves (by B. Cosgrove, 9 November 1992), Capricornia Collection, CQU.

Bibliography

Transcript of interview with Colin and Vi Heberlein (by B. Cosgrove, 21 October 1992), Capricornia Collection, CQU.
Interview with Jim Leigh (by Betty Cosgrove, 15 September 1992), Capricornia Collection, CQU.
Transcript of interview with Lorna McDonald (by B. Cosgrove, 10 and 19 August 1992), Capricornia Collection, CQU.
Interview with Lawrence Miller (by Alex Hood, 29 May 2002), Oral TRC 4864/4–5, NLA.
Interview with Sir Frank Nicklin (by Suzanne Lunney, 19 March 1974), Oral TRC 254, NLA.
Interview with Paul Said (by Barry York, 12 September 1984), Oral TRC 3582/2, NLA.
Transcript of interview with Christine Schluter (by B. Cosgrove, 5 November 1992), Capricornia Collection, CQU.
Interview with Charlie Shannon (by Betty Cosgrove, 10 September 1992), Capricornia Collection, CQU.
Transcript of interview with Trevor James Stock (by Lorna McDonald, 24 November 1992) Capricornia Collection, CQU.
Transcript of interview with Alex Teeny (by B. Cosgrove, 8 October 1992), Capricornia Collection, CQU.
Interview with Max Thompson (by John Merritt, 5 December 2001), Oral TRC 4822, NLA.
Interview with Arthur Timms (by Ray Aitchison, 14 August 1986), Oral TRC 2061, NLA.
Transcript of interview with Bill Toby (by B. Cosgrove, 29 April and 5 May 1993), Capricornia Collection, CQU.
Transcript of interview with Sydney 'Syd' Wall (by B. Cosgrove, 10 November 1992), Capricornia Collection, CQU.
Interview with Leonard Zubrinich (by Edward Stokes in the Broken Hill social history project, 26 June 1982), NLA.

Secondary sources

Books

Australian Commodity Statistics, 1995. ABARE, Canberra, 1995.
Australia's Identified Mineral Resources, 2008. Geoscience Australia/Australian Government, Canberra, 2008.
Adams, Christine. *Sharing the Lode: The Broken Hill migrant story*. Broken Hill Migrant Heritage Committee, Broken Hill, 2004.
Adams, Paul Robert. *The Best Hated Man in Australia: The Life and Death of Percy Brookfield, 1875–1921*. Puncher & Wattmann, Sydney, 2010.
Alpin, G. Glynn, S. and McKernan. M. (eds). *Australians: Events and places*. Fairfax, Syme & Weldon Associates, Sydney, 1987.
Andersen, Lisa and Andrew, Jane. *Quality of Light, Quality of Life: Professional artists and cultural industries in and around Broken Hill*. Regional Arts NSW, Sydney, 2007.
Australian UNESCO Seminar, *Man and the Environment: New towns in isolated settings*. Australian Government Publishing Service, Canberra, 1976.

Australia's Identified Mineral Resources, 2008. Geoscience Australia, Australian Government, Canberra, 2008.

Attwood, Bain. *The Making of the Aborigines*. Allen & Unwin, Sydney, 1989.

Australian Encyclopedia, Vol. 2 (4th edn). Grolier Society, Sydney, 1983.

Banks, F.E. *Copper in the World Economy*. Ballinger Publishing Company, Cambridge MA, 1974.

Banks, M.R. and Kirkpatrick, J.B. (eds). *Landscape and Man: The interaction between man and environment in Western Tasmania*. Proceedings of a Symposium Organized by the Royal Society of Tasmania. Royal Society of Tasmania, Hobart, June 1977.

Bauman, Zygmunt. *Liquid Modernity*. Polity Press, Cambridge, 2000.

Berkman, Don. *Making the Mount Isa Mine, 1923–1933*. The Australasian Institute of Mining and Metallurgy, Spectrum Series No. 1, Melbourne, 1996.

Blainey, Geoffrey. *The Peaks of Lyell* (5th edn). St David's Park Publishing, Hobart, 1993.

Blainey, Geoffrey. *The Rise of Broken Hill*. Macmillan, Melbourne, 1968.

Blainey, Geoffrey. *The Rush That Never Ended: A history of Australian mining*. MUP, Melbourne, 1963.

Blainey, Geoffrey. *The Steel Master: A life of Essington Lewis*. Macmillan, Melbourne, 1971.

Blainey, Geoffrey. *The Tyranny of Distance: How distance shaped Australia's history* (revised edn). Macmillan, Melbourne, 1982.

Bolton, Geoffrey. *A Thousand Miles Away: A history of North Queensland to 1920*. Jacaranda, ANU, Canberra, 1963.

Bolton, Geoffrey. *Spoils and Spoilers: Australians make their environment, 1788–1980* (2nd edn). Allen & Unwin, Sydney, 1992.

Borchardt, D.H. and Crittenden, Victor (eds). *Australians: A guide to sources*. Fairfax, Syme & Weldon, Sydney, 1987.

Bottom, Bob. *Behind the Barrier*. Gareth Powell Associates, Hong Kong, 1969.

Bowman, Margaret. *Beyond the City: Case studies in community structure and development*. Longman Cheshire, Melbourne, 1981.

Boyce, James. *Van Diemen's Land*. Black Inc, Melbourne, 2008.

Bridges, Roy. *From Silver to Steel: The romance of the Broken Hill Proprietary*. Robertson, Melbourne, 1920.

Buckley, Ken and Wheelwright, Ted (eds). *Essays in the Political Economy of Australian Capitalism*, Vol. 5. Australian New Zealand Book Company, Sydney, 1983.

Bullock, Ken. *Port Pirie: The friendly city – the undaunted years*. Peacock Publications, Norwood (SA), 1988.

Burke, Graham. *Football Heroes of the Hill: Those wild old westies*. Graham Burke, Melbourne, 1999.

Buxton, Geoffrey. *The Riverina, 1861–1891: An Australian regional study*. MUP, Sydney, 1967.

Cain, Frank. *The Wobblies at War: A history of the IWW and the Great War in Australia*. Spectrum, Melbourne, 1993.

Camilleri, Jenny. *In the Broken Hill Paddock*. Jenny Camilleri, Broken Hill, 2006.

Camilleri, Jenny. *Some Outstanding Women of the Broken Hill and District*. Jenny Camilleri, Broken Hill, 2002.

Carolan, Jayne Mayo. *No Run-of-the-Mill: A biography of Henry Beaufort Somerset*. Australian Scholarly Publishing, Melbourne, 2006.

Chaput, Donald and Kennedy, Kett. *The Man from ASARCO: A life and times of Julius Kruttschnitt*. Australasian Mineral Heritage Trust/Australasian Institute of Mining

Bibliography

and Metallurgy, Melbourne, 1992.
Chardon, N. and Golding, F.L. (eds). *Centenary of the Town of Mount Morgan, 1882–1982*. Mount Morgan and District Historical Society, Mount Morgan, 1982.
Clark, John (ed.). *Australia in Maps: Great maps in Australia's history from the National Library's Collection*. National Library of Australia, Canberra, 2007.
Clark, Lindesay. *Built on Gold: Recollections of Western Mining*. Hill of Content, Melbourne, 1983
Cilento, Raphael and Lack, Clem. *Triumph in the Tropics: An historical sketch of Queensland*. The Historical Committee of The Centenary Celebrations Council of Queensland, Brisbane, 1959.
Cochrane, Peter. *Industrialisation and Dependence: Australia's road to economic development*. UQP, Brisbane, 1980.
Cohen, A.P. *The Symbolic Construction of Community*. Routledge, London, 1989.
Conlon, Ann and Ryan, Edna. *Gentle Invaders: Australian women at work* (2nd edn). Penguin, Melbourne, 1989.
Connell R.W. and Irving, T.H. *Class Structure in Australian History: Poverty and progress*. Longman Cheshire, Melbourne, 1992.
Consolidated Zinc Corporation Limited. Consolidated Zinc Corporation, London, 1960.
Cook, Kenneth. *Wake in Fright*. St Martin's Press, New York, 1962.
Cooke, Philip. *Back to the Future: Modernity, post modernity and locality*. Unwin Hyman, London, 1990.
35 Years of the Q.C.W.A in Mount Isa, 1928–1963. Mount Isa, Queensland Country Women's Association Branch, 1963.
Cowie, William Gordon. *Some Account of the Mount Morgan Gold Mine, Rockhampton, Queensland*. Morning Bulletin, Rockhampton, 1885.
Curthoys, Ann, Martin, A.W. and Rowse, Tim (eds). *Australians from 1939*. Fairfax, Syme & Weldon, Sydney, 1987.
Curtis, Leonard Samuel (compiler and editor). *The History of Broken Hill: Its rise and progress*. Frearson's Printing House, Adelaide, 1908.
Dale, George. *The Industrial History of Broken Hill*. Fraser & Jenkinson, Melbourne, 1918.
Daniels, Kay. *Convict Women*. Allen & Unwin, Sydney, 1998.
Dempsey, Frank. *Old Mining Towns of North Queensland*. Rigby, Adelaide, 1980.
Dempsey, Ken. *Smalltown: A study of social cohesion and belonging*. OUP, Melbourne, 1991.
Dick, W.H. *The Famous Mount Morgan Gold Mine, Rockhampton, Queensland: Its history, geological formation and prospects*. William Munro, Rockhampton, 1888.
Dixon, Robert and Kelly, Veronica (eds). *Impact of the Modern: Vernacular modernities in Australia, 1870s–1960s*. Sydney University Press, Sydney, 2008.
Docherty, J.C. *Newcastle: The making of an Australian city*. Hale & Iremonger, Sydney, 1983.
Donley, R.J.R. *The Rise of Port Pirie*. R.J.R. Donley, Adelaide, 1975.
Douglas, Louise. *Oral History: A handbook*. Allen & Unwin, Sydney, 1988.
Edwards, Richard C. *Contested Terrain: The transformation of work in the twentieth century*. Basic Books, New York, 1979.
Eklund, E. *Steel Town: The making and breaking of Port Kembla*. MUP, Melbourne, 2002.
Eklund, E. and Murray, M. *Copper Community: The Electrolytic Refining and Smelting Company of Australia Ltd & Southern Copper Ltd, Port Kembla*. University of Wollongong Press, Wollongong, 2000.
Elphick, Beverley and Elphick, Don (compilers and editors). *Menindee Mission Station*,

1933–1949. B. & D. Elphick, Canberra, 1996.
Evans, Ray. *A History of Queensland*. CUP, Cambridge, 2007.
Farwell, George. *Down Argent Street: The story of Broken Hill* (illustrated with drawings by Roy Dalgarno). F.H. Johnston Publishing Co. Pty Limited, Sydney, 1948.
Fitzgerald, Shirley. *Rising Damp: Sydney, 1870–1890*. OUP, Melbourne, 1987.
Forster, Colin. *Industrial Development in Australia, 1920–1930*. ANU Press, Canberra, 1964.
Frisch, Michael. *Shared Authority: Essays on the craft and meaning of oral and public history*. State University of New York Press, New York, 1990.
Gammage, Bill and Markus, Andrew. *All that Dirt: Aborigines, 1938*. An Australian 1938 Monograph, Canberra, 1982.
Gammage, Bill and Spearritt, Peter (eds). *Australians 1938*. Fairfax, Syme & Weldon, Sydney, 1987.
Gardner, M. and Palmer, G. *Employment Relations: Industrial relations and human resource management in Australia*. Macmillan, Melbourne, 1992.
Glynn, Sean. *Urbanisation in Australian History* (3rd edn). Nelson, Melbourne, 1975.
Goldberg, S.L. and Smith, F.B. (eds). *Australian Cultural History*. CUP, Cambridge, 1988.
Gollan, Robin. *Radical and Working Class Politics: A study of eastern Australia 1850–1910*. MUP, Melbourne, 1960.
Gollan, Robin. *Revolutionaries and Reformists: Communism and the Australian labour movement, 1920–1950*. Allen & Unwin, Sydney, 1975.
Gollan, Robin. *The Coalminers of New South Wales: A history of the union, 1860–1960*. MUP, Melbourne, 1963.
Gostin, V.A. (ed.). *Gondwana to Greenhouse: Australian environmental geoscience*. Geological Society of Australia Inc., Special Publication No. 21, Southwood Press, Sydney, 2001.
Grabs, Cyril. *Gold, Black Gold and Intrigue: The story of Mount Morgan* (2nd edn). Central Queensland University Press, Rockhampton, 2000.
Gray, Ian. *Politics in Place: Social power relations in an Australian country town*. CUP, Cambridge, 1991.
Green, David and Cromwell, Lawrence. *Mutual Aid or Welfare State: Australia's Friendly Societies*. Allen & Unwin, Sydney, 1984.
Green, F.A. *The Port Pirie Smelters*. Broken Hill Associated Smelters, Melbourne, 1977.
Gregory, David and Urry, John (eds). *Social Relations and Spatial Structures*. Macmillan, Hampshire (UK), 1985.
Gresham, J.J. *Kambalda: History of a mining town*. Western Mining Corporation Limited, Melbourne, 1991.
Hagan, Jim (ed.). *People and Politics in Regional New South Wales, Vol. 2: The 1950s to 2006*. The Federation Press, Sydney, 2006.
Hancock, W.K. *Australia* (reprinted with a new preface), Jacaranda Press, Brisbane, 1961 [1930].
Hancock, W.K. *Discovering Monaro: A study of man's impact on his environment*. CUP, Cambridge, 1972.
Hardy, Bobbie. *Water Carts to Pipelines: The history of the Broken Hill water supply*. Broken Hill Water Board/Halstead Press, Sydney, 1968.
Harvey, David. *Limits of Capital*. Basil Blackwell, Oxford, 1982.
Harvey, David. *The Condition of Post Modernity: An enquiry into the origins of cultural change*. Basil Blackwell, Oxford, 1989.

Bibliography

Hearn, Mark and Knowles, Harry. *One Big Union: A history of the Australian Workers Union 1886–1994*. CUP, Cambridge, 1996.

Hercus, L.A. *A Nukunu Dictionary*. L.A. Hercus, Canberra, 1992.

Hirst, J.B. *Adelaide and the Country, 1870–1917: Their social and political relationship*. MUP, Melbourne, 1973.

Hore-Lacy, Ian (ed.). *Broken Hill to Mount Isa: The mining odyssey of W.H. Corbould*. Hyland House, Melbourne, 1981.

Howard, W.A. *Barrier Bulwark: The life and times of Shorty O'Neil*. Willry Pty Ltd, Melbourne, 1990.

Huf, L., McDonald, L. and Myers, D. (eds). *Sin, Sweat and Sorrow: The making of Capricornia, Queensland, 1840–1940s*. UQP, Rockhampton, 1993.

Hughes, C.A. and Graham, B.D. *A Handbook of Australian Government and Politics, 1890–1964*. ANU Press, Canberra, 1968.

Huxley, Elspeth. *Their Shining Eldorado: A journey through Australia*. Chatto & Windus, London, 1967.

Idriess, Ion. *The Silver City*. Angus & Robertson, Sydney, 1956.

Iremonger, John, Merritt, John and Osborne, Graeme (eds). *Strikes: Studies in twentieth century Australian social history*. Angus & Robertson in association with the Australian Society for the Study of Labour History, Sydney, 1973.

Jones, Rees Rutland. *Gold Mining in Central Queensland and the Mount Morgan Mine*. Morning Argus, Rockhampton, 1913.

Kearns, R.H.B. *Broken Hill, Vol. 4: The First Century, 1940–1983* (revised edn). Broken Hill Historical Society, Broken Hill, 1987.

Kennedy, Brian. *Silver, Sin and Six Penny Ale: A social history of Broken Hill*. MUP, Melbourne, 1978.

Kerr, John. *Mount Morgan: Gold, copper and oil*. J.D & R.S. Kerr, Brisbane, 1982.

King, Norma. *Nickel Country – Gold Country*. Rigby, Adelaide, 1972.

Kingston, B. *The Oxford History of Australia, Vol. 3, 1860–1942, Glad, Confident Morning*. OUP, Melbourne, 1988.

Kirkman, N. *Mount Isa: Oasis of the outback*, James Cook University, Townsville, 1998.

Knibbs, G.H. *Local Government in Australia*. Commonwealth Bureau of Census and Statistics, Melbourne, 1919.

Kriegler, Roy. *Working For The Company: Work and control in the Whyalla shipyard*. OUP, Melbourne, 1980.

Lines, Peter (compiler). *Encyclopaedia of South Australian Country Football Clubs*. Peter Lines, Cowell (SA), 2008.

Lines, William J. *Taming the Great South Land: A history of the conquest of nature in Australia*. University of California Press, Berkeley, 1991.

Long, J.W. *The Early History of Mt Morgan, Dawson and Callide Valleys*. [Mt Morgan?], 1962.

Love, Peter. *Labour and the Money Power: Australian labour populism, 1890–1950*. MUP, Melbourne, 1984.

Lowenstein, Wendy. *Weevils in the Flour: An oral record of the 1930s Depression in Australia*. Scribe, Melbourne, 1981.

Lowenthal, David. *The Past is a Foreign Country*. CUP, Cambridge, 1985.

Lundager, J.H. *The Early History of Rockhampton*. Rockhampton Morning Bulletin, Rockhampton, 1909.

Macintyre, Stuart. *The Oxford History of Australia, Vol. 4, 1901–1942, The Succeeding Age*. OUP, Melbourne, 1986.

Mackie, Pat (with Elizabeth Vassilieff). *Mount Isa: The story of a dispute.* Hudson, Hawthorn (Vic), 1989.
McCalman, Janet. *Struggletown: Portrait of an Australian working-class community.* Penguin, Melbourne, 1984.
McCarthy, Ken. *Steaming Down Argent Street: A history of the Broken Hill steam tramways 1902–1926.* The Sydney Tramway Museum, Sydney, 1983.
McDonald, Lorna. *Rockhampton: A history of city and district.* UQP, Brisbane, 1980.
McFarlane, Ian. *Beyond Awakening: The Aboriginal tribes of north west Tasmania: A History.* Fullers Bookshop/Riawunna/University of Tasmania, Launceston (Tas), 2008.
McGowan, Barry. *Dust and Dreams: Mining communities in south-east New South Wales.* UNSW Press, Sydney, 2010.
McGuire, Paul. *Australian Journey* (new and revised edn). William Heinemann, London, 1942.
McIntyre, A.J. and J.J. *Country Towns of Victoria: A social survey.* MUP, Melbourne, 1944.
McLean, Donald. *Broken Hill Sketchbook* (illustrations by Frank Beck). Rigby, Adelaide, 1968.
McNally, Ward. *To Broken Hill and Back.* Widescope International, Camberwell (Vic), 1975.
Massey, Doreen. *The Spatial Divisions of Labour.* Macmillan, London, 1984.
Maxwell-Stewart, Hamish. *Closing Hell's Gates: The death of a convict station.* Allen & Unwin, Sydney, 2008.
Mayne, Alan (ed.). *Beyond the Black Stump: Histories of outback Australia.* Wakefield Press, Adelaide, 2006.
Mayne, Alan and Atkinson, Stephen (ed.). *Outside Country: A history of inland Australia.* Wakefield Press, Adelaide, 2011.
Meinig, D.W. *On the Margins of the Good Earth: The South Australian wheat frontier, 1869–1884.* South Australian Government Printer, Netley (SA), 1988.
Meredith, David and Dyster, Barrie. *Australia in the Global Economy: Continuity and change.* CUP, New York and Melbourne, 1999.
Merritt, John and Osborne, Graeme (eds). *Strikes: Studies in twentieth century social history.* Angus & Robertson in association with the Australian Society for the Study of Labour History, Sydney, 1973.
Moss, Jim. *The Sound of Trumpets: History of the labour movement in South Australia.* Wakefield Press, Gondawilla (SA), 1985.
Mount Isa Mine Ltd. MIM Ltd, Mount Isa, 1939.
Mount Isa Today. MIM Ltd, Mount Isa, 1967.
Murphy, D.J. (ed.). *Labor in Politics: The state labor parties in Australia, 1880–1920.* UQP, Brisbane, 1975.
Murphy, D.J. (ed.). *The Big Strikes: Queensland, 1889–1965.* UQP, Brisbane, 1983.
Nairn, Bede and Serle, Geoffrey (eds). *The Australian Dictionary of Biography, Vol. 8: 1891–1939.* MUP, Melbourne, 1981.
Neal, C., Tykklainen M. and Bardbury, J. (eds). *Coping with Closure: An international comparison of mine town experiences.* Routledge, London, 1992.
Newman, J. Malcolm. *Guide to the Operations of Mount Morgan Limited.* Mount Morgan, 1960.
Newtown, R. *The Work and Wealth of Queensland.* Outride & Co., Brisbane, 1897.
Nisbet, Hume. *A Colonial Tramp: Travels and adventures in Australia and New Guinea.* Ward & Downey, 1891.
O'Connor, Desmond (ed). *Memories and Identities: Proceedings of the second conference on the*

Bibliography

impact of Italians in South Australia. Australian Humanities Press, Adelaide, 2004.
O'Connor, Desmond. *No Need To Be Afraid: Italian settlers in South Australia between 1939 and the Second World War*. Wakefield Press, Adelaide, 1996
O'Connor, Desmond and Comin, Antonio (eds). *The First Conference on the Impact of Italians in South Australia (16–17 July, 1993): Proceedings*. Italian Congress Inc. and Flinders University, Adelaide, 1993.
Official Souvenir of Port Pirie Diamond Jubilee, 1876–1936. Port Pirie Town Council, 1936.
O'Neil, Bernard, Raftery, Judith and Round, Kerrie (eds). *Playford's South Australia: Essays on the history of South Australia, 1933–1968*. Association of Professional Historians, Adelaide, 1996.
Operations of Mount Morgan Limited, Mount Morgan, Queensland (7th edn). Mount Morgan, 1969.
Queensland: Past and Present: 100 Years of Statistics, 1896–1996. Office of Economic and Statistical Research, Queensland Government, Brisbane, 1996.
Quin's Metal Handbook and Statistics. Metal Information Bureau, London, 1953.
Ralph, G.M. *A Pictorial History of Kambalda*. Western Mining Corporation, Melbourne, 1992.
Reeves, A. *Another Day Another Dollar: Working class lives in Australian history*. McCulloch Publishing, Melbourne, 1988.
Reid, Don. *They Searched: A History of Exploration within Western Mining Corporation, Part 1 – Sawdust and Ice: The First Twenty Years*. Western Mining Corporation, Melbourne, 1981.
Reiger, Kerreen. *The Disenchantment of the Home: Modernising the Australian family, 1880–1940*. OUP, Melbourne, 1988.
Reynolds, Henry. *Dispossession: Black Australians and white invaders*. Allen & Unwin, Sydney, 1989.
Reynolds, Henry. *Fate of a Free People*. Penguin, Melbourne, 1995.
Reynolds, Henry. *Frontier, Aborigines, Settlers and Land*. Allen & Unwin, Sydney, 1987.
Reynolds, Henry. *A History of Tasmania*. CUP, Melbourne, 2011.
Reynolds, Henry. *The Other Side of the Frontier*. Penguin, Melbourne, 1982.
Rich, D.C. *The Industrial Geography of Australia*. Methuen, Sydney, 1986.
Rickard, John. *Class and Politics: New South Wales, Victoria and the Early Commonwealth, 1890–1910*. ANU Press, Canberra, 1976.
Rintoul, John. *Esperance: Yesterday and today* (4th edn). Esperance Shire Council, Perth, 1986.
Ritchie, John (ed.). *Australian Dictionary of National Biography, 1891–1939*. MUP, Melbourne, 1990.
Robinson, Nancy. *The Reluctant Harbour: The romance of Pirie*. Nadjuri Australia, Jamestown (SA), 1976.
Robinson, W.S. *If I Remember Rightly: The memoirs of W.S. Robinson* (ed. Geoffrey Blainey), Cheshire, Melbourne, 1967.
Rowley, C.D. *The Destruction of Aboriginal Society, Aboriginal Policy and Practice, Vol. 1*. ANU Press, Canberra, 1970.
Rowse, Tim. *Australian Liberalism and National Character*. Kibble, Malmsbury (Vic), 1978.
Ryan, Lyndall. *The Aboriginal Tasmanians* (2nd edn). Allen & Unwin, Sydney, 1996.
Samuel, Raphael (ed.). *People's History and Socialist Theory*. Routledge Kegan Paul, London, 1981.
Shanks A.G. and Boyle, R.F. *The First Hundred Years: Mount Morgan Lodge No. 57, United Grand Lodge of Queensland*. Mount Morgan Lodge No. 57, Mount Morgan, 1988.

Shaw, A.G.L. *The Story of Australia*. Faber & Faber, London, 1955.
Shaw, A.G.L. and Clark, C.M.H. *Australian Dictionary of Biography, Vol. 1, 1788–1850*. MUP, Melbourne, 1967.
Sheldon, Gordon. *Industrial Siege*. Cheshire, Melbourne, 1965
Shields, John (ed.). *All Our Labours: Oral histories of working life in twentieth century Sydney*. UNSW Press, Sydney, 1992.
Skivinis J. and Taylor, A. *A Mining Town: Kambalda*. Longman, Melbourne, 1975.
Solomon, R.J. *The Richest Lode: Broken Hill, 1883–1988*. Hale & Iremonger, Sydney, 1988.
Southerwood, W.T. *Planting a faith in Queenstown*. The Catholic Church, Queenstown, 1953.
Spearritt, Peter and Walker, David. *Australian Popular Culture*. Allen & Unwin, Sydney, 1979.
Spearritt, Peter. *Sydney Since the Twenties*. Hale & Iremonger, Sydney, 1978.
St Andrew's, 'the friendly church', Queenstown, Tasmania. St Andrew's, Queenstown, 1956.
Stannage, T. (ed.). *A New History of Western Australia*. University of Western Australia (UWA) Press, Perth, 1981.
Stannage, T. *The People of Perth: A social history of Western Australia's capital*. UWA Press, Perth, 1979.
Sturt, Charles. *Narrative of An Expedition performed under the authority of Her Majesty's Government, during the years 1844, 5, and 6: together with a notice of the province of South Australia, in 1847, Vol. 2*. T. & W. Bone, London, 1849.
Svensen, Stuart. *Industrial War: The great strikes, 1890–1894*. Ram Press, Wollongong (NSW), 1995.
Sykes, Trevor. *The Money Miners: The great Australian boom, 1969–1970*. Wildcat Press, Sydney, 1978.
The Australia Directory (10th edn). Hydrographic Office, British Admiralty.
The Lead and Zinc Industry at Port Pirie, South Australia (new edn). Broken Hill Associated Smelters, Melbourne, 1982.
The Silver-Lead Mines of Broken Hill NSW – Death and Disease. Broken Hill Branch Communist Party, Broken Hill (NSW), 1933?
The Story of the Mount Morgan Mine, 1882–1957. Mount Morgan Ltd, Mount Morgan, 1957.
Thompson, E.P. *The Making of the English Working Class* (2nd edn). Pelican, Harmondsworth (UK), 1968.
Thompson, E.P. *The Poverty of Theory and Other Essays*. Merlin Press, London, 1978.
Thompson, Paul. *The Voice of the Past: Oral history*. OUP, Oxford, 1978.
Thrift, Nigel and Williams, Paul (eds). *Class and Space: The making of urban society*. Routledge & Kegan Paul, London, 1988.
Townsley, W.A. *Tasmania: From colony to statehood*. St David's Park Publishing, Hobart, 1991.
Trigger, D. *Whitefella Comin': Aboriginal responses to colonialism in northern Australia*. CUP, Cambridge, 1992.
Trigger, D. and Griffiths, G. (eds). *Disputed Territories: Land, culture and identity in settler societies*. Hong Kong University Press, Hong Kong, 2003.
Upfield, Arthur. *The Bachelors of Broken Hill*. Heinemann, London, 1958.
Vamplew, W. (ed.). *Australians: Historical statistics*. Fairfax, Syme & Weldon, Sydney, 1987.
Wanna, John. *Defence not Defiance: The development of organised labour in South Australia*. Texts in Humanities, Adelaide, 1981.

Bibliography

Ward, Russell. *The Australian Legend* (2nd edn). OUP, Melbourne, 1966.
Waterson, D.B. *Squatter, Selector and Storekeeper: A history of the Darling Downs, 1859–93*. Sydney University Press, Sydney, 1968.
Webber, Horace. *The Greening of the Hill: Re-vegetation around Broken Hill in the 1930s*. Hyland House, Melbourne, 1992.
White, Neil. *Company Towns: Corporate order and community*. University of Toronto Press, Toronto (Canada), 2011.
White, Richard. *Inventing Australia: Images and identity, 1688–1980*. Allen & Unwin, Sydney, 1981.
Wild, R.A. *Australian Community Studies and Beyond*. Allen & Unwin, Sydney, 1981.
Williams, Michael. *The Making of the South Australian Landscape: A study of the historical geography of Australia*. Academic Press, London, 1974.
Woodman, E.F. *The Catholic Church in Broken Hill, 1883 – 1983: The first hundred years*. E.F. Woodman, Broken Hill (NSW), 1983.
Wright, C. *The Management of Labour: A history of Australian employers*. MUP, Melbourne, 1995.

Articles and reports

Adams, Paul Robert and Eklund, Erik. 'Representing Militancy: Photographs of the Broken Hill industrial disputes, 1908–1920'. *Labour History*, no. 101, November 2011, pp. 1–34.
Beckett, Jeremy. 'Marginal Men: A study of two half-caste [sic] Aborigines'. *Oceania*, vol. 29, no. 2, 1958, pp. 91–108.
Blackmur, Douglas. 'Arbitration, Legislation and Industrial Peace: Queensland in the reconstruction years'. *Labour History*, no. 63, November 1992, pp. 115–34.
Boyd, Adam. 'A History of Mount Morgan'. *Australasian Institute of Mining and Metallurgy, Proceedings, New Series*, no. 115, 1939.
Boyle R.F. and Gistitin, C. 'Mount Morgan – A New Beginning'. *Proceedings of the Sixth National Conference on Engineering Heritage*, Hobart, 5–7 October 1992.
Broken Hill, NSW, Heritage and Cultural Tourism Program, 1986–2005 Productivity commission submission July 2005. Broken Hill City Council, Broken Hill, 2005.
Caprioni, E., Deiana, M. and Eklund, E. 'Interview Techniques in Three Different Research Scenarios'. *The International Journal of Interdisciplinary Social Sciences*, vol. 4, no. 1, 2009, pp. 1–10.
Bretta, Carthey and Potter, Bob. *Mount Isa: the Great Queensland strike*, Solidarity Pamphlet No. 22. < libcom.org/library/mount-isa-great-queensland-strike-solidarity>.
Cartwright, B., Merry, R.H. and Tiller, K.G. 'Heavy metal contamination of soils around a lead smelter at Port Pirie, South Australia'. *Australian Journal of Soil Research*, vol. 15, no. 1, 1976, pp. 69–81.
'Copper Mining at Mount Lyell', Reprint from the *Chemical Engineering and Mining Review*, November 1940–February 1941,Tait Publishing Company, Melbourne, 1941?, pp. 4–7.
Corbet, K.D. 'New Mapping and Interpretations of the Mount Lyell Mining District, Tasmania: A large hybrid Cu-Au system with an exhalative Pb-Zn top'. *Economic Geology*, vol. 96, 2001, pp. 1089–122.
Curthoys, Ann and Hamilton, Paula. 'What Makes History Public?' *Public History Review*, vol. 1, 1992, pp. 8–13.
Davies, P. Mitchell, N. and Barmuta, L. *The impact of historical mining operations at Mount*

Lyell on the water quality and biological health of the King and Queen River catchments, Western Tasmania. Supervising Scientist Report No. 118, 1996.

Davison, Graeme. 'The Uses and Abuses of Australian History'. *Australian Historical Studies*, vol. 23, no. 91, 1988, pp. 55–76.

Deacon, Desley. 'Taylorism in the Home'. *Australian and New Zealand Journal of Sociology*, vol. 21, no. 2, 1985, pp. 161–73.

Duncan, Stuart and Savage, Martin. 'Commentary: New perspectives on the locality debate'. *Environment and Planning A*, vol. 23, 1991, pp. 155–64.

Eklund, Antoinette and Eklund, Erik. 'Do you Love the Town you Live in?': Narratives of place from Australian mining towns'. *The International Journal of Interdisciplinary Social Sciences*, vol. 3, no. 7, 2008, pp. 53–57.

Eklund, Erik. 'We Are of Age: Class, locality and region at Port Kembla, 1900 to 1940'. *Labour History*, no. 66, May 1994, pp. 72–85.

Eklund, Erik. 'Managers, workers and industrial welfarism: Management strategies at ER&S, Port Kembla and the Sulphide Corporation, Cockle Creek, 1895–1929'. *Australian Economic History Review*, vol. 37, no. 2, 1997, pp. 137–57.

Eklund, Erik. 'Official and Vernacular Public History: Historical anniversaries and commemorations in Newcastle, NSW'. *Public History Review*, vol. 14, 2007, pp. 128–52.

Eklund, Erik. 'Intelligently Directed Welfare Work?: Labour management strategies in local context: Port Pirie, 1915–29'. *Labour History*, vol. 76, May 1999, pp. 125–48.

Eklund, Erik. 'The "place" of politics: Class and localist politics at Port Kembla, 1900–1930', *Labour History*, no. 78, May 2000, pp. 94–115.

Ellem, Bradon and Shields, John. 'Making a "Union Town": Class, Gender and Consumption in Inter-War Broken Hill'. *Labour History*, no. 78, May 2000, pp. 116–40.

Ellem, Bradon and Shields, John. 'Making the "Gibraltar of Unionism": Union organising and peak union agency in Broken Hill, 1886–1930'. *Labour History*, vol. 83, November 2002, pp. 65–88.

Ellem, Bradon and Shields, John. 'Why do Unions form Peak Bodies? The case of the Barrier Industrial Council'. *Journal of Industrial Relations*, vol. 38. no. 3, 1996, pp. 377–411.

Ford, A. and Blenkinsop, T.G. 'Combining fractal analysis of mineral deposit clustering with weights of evidence to evaluate patterns of mineralization: Application to copper deposits of the Mount Isa Inlier, NW Queensland, Australia'. *Ore Geology Review*, vol. 33, 2008, pp. 435–50.

Forster, Colin. 'Australian Manufacturing and the War of 1914–18'. *Economic Record*, vol. 29, November 1953, pp. 211–30.

Fox, Charlie. 'Contract Work at Mount Lyell'. *Journal of Australasian Mining History*, vol. 6, September 2008, pp. 90–110.

Fox, Charlie. 'Work and Welfare at Mt Lyell, 1913–1923'. *Journal of Australasian Mining History*, vol. 1, no. 1, 2003, pp. 1–78.

French, Maurice. 'Local History and a Regional College: The Darling Downs experience'. *Australian Historical Association Bulletin*, no. 23, July 1980, pp. 5–13.

Gaunt, Heather. 'The Library of Robert Carl Sticht'. *La Trobe Journal*, no. 79, Autumn 2007, pp. 4–26.

Gibson, Katherine. 'Considerations on Northern Marxist Geography: A review from the Antipodes'. *Australian Geography*, vol. 22, no. 1, 1991, pp. 75–81.

Gillespie, R. 'Accounting for lead poisoning: The medical politics of occupational

health'. *Social History*, vol. 15, no. 3, 1990, pp. 303–32.
Goad, Philip. 'Mary Kathleen and Weipa: Two model townships for post-war Australia'. *Transition*, vol. 49/50, 1996, pp. 44–59.
Gregson, Sarah. 'Defending Internationalism in Interwar Broken Hill'. *Labour History*, no. 86, May 2004, pp. 115–36.
Gistitin C. and Boyle, R.F. 'The Golden Mount – Central Queensland's First Triumph'. *Proceedings of the Fifth National Conference on Engineering Heritage*, Perth, 3–5 December 1990.
Griffin, Grahame. 'J.H.Lundager, Mount Morgan Politician and Photographer: Company hack or subtle subversive?'. *Journal of Australian Studies*, no. 34, September 1992, pp. 15–31.
Hagan J. and Castle, R. 'Regulation of Aboriginal labour in Queensland: Protectors, agreements and trust accounts, 1897–1965'. *Labour History*, no. 72, May 1997, pp. 66–76.
Harvey, David. 'Spatial Fix and Globalization', *Geographische Revue*, vol. 2, 2001, pp. 23–30.
Hope, Jeannette. 'Unincorporated Area of NSW – A Heritage Study', A Report for the Department of Natural Resources and the Heritage Office of NSW, River Junction Research, 2006.
Hoatson, Dean M., Jaireth, Subhash Jaques and Lynton, A. 'Nickel sulfide deposits in Australia: Characteristics, resources, and potential'. *Ore Geology Reviews*, vol. 29, issues 3–4, 2006, pp. 177–241.
Jones, Rhys. 'Tasmanian Archaeology: Establishing the Sequences'. *Annual Review of Anthropology*, vol. 24, 1995, pp. 423–46.
Keele, Svenja, 'Lunar Landscapes and sulphurous smogs: An environmental history of human impacts on the Queenstown region in western Tasmania c. 1890–2003'. Research Paper No. 18, School of Anthropology, Geography and Environmental Studies, University of Melbourne, 2003.
Keith Johns, R. 'Uranium in South Australia – Politics and Reality'. *Journal of Australasian Mining History*, vol. 3, September 2005, pp. 171–84.
Kennedy, Brian. 'Regionalism and Nationalism: Broken Hill in the 1880s'. *Australian Economic History Review*, vol. 20, no. 1, 1980, pp. 64–76.
Kimber, Julie. 'A Case of Mild Anarchy?: Job Committees in the Broken Hill mines, c.1930 to c.1954'. *Labour History*, no. 80, May 2001, pp. 41–64.
Kirkpatrick, R. 'How a communist rag in Darwin became a newspaper for Mount Isa'. *PANPA Bulletin*, June 2005, pp. 56–57.
Kanowna Heritage Trail. Heritage Council of Western Australia/Delta, Gold N.L., 1993.
Koehnken, L. *Final Report: Mount Lyell Remediation Research and Demonstration Program*. Supervising Scientist Report No. 126, Canberra, 1997.
Laite, Julianne. 'Historical Perspectives on Industrial Development, Mining and Prostitution'. *The History Journal*, vol. 52, issue 3, 2009, pp. 739–62.
Latana, Masterman & Associates, *A Tale of Tin and Silver: Broken Hill heritage study*. Latana, Masterman & Associates, Sydney, 1987.
Lee, C. and Stabin-Nesmith, B. 'The Continuing Value of a Planned Community: Radburn in the evolution of suburban development'. *Journal of Urban Design*, vol. 6, no. 2, 2001, pp. 151–84.
Lee, Jenny and Fahey, Charles. 'A Boom for Whom?: Some developments in the Australian labour market, 1870–1891'. *Labour History*, no. 50, May 1986, pp. 1–27.
Lester Firth & Murton Pty Ltd, *Port Pirie Conservation Study*. Department of Urban and

Regional Affairs, Adelaide, 1980.
Lipton, Phillip. 'A History of Company Law in Colonial Australia: Legal evolution and economic development'. Corporate Law and Accountability Research Group, Working Paper No. 10, Department of Business Law and Taxation, Monash University, 2007.
Littman, W. 'Designing Obedience: The architecture and landscape of welfare capitalism, 1880–1930'. *International Labor and Working Class History*, no. 53, Spring 1998, pp. 88–114.
Lottermoser, B.G., Ashley, P.M. and Costelloe, M.T. 'Contaminant dispersion at the rehabilitated Mary Kathleen uranium mine, Australia'. *Environmental Geology*, vol. 48, no. 6, 2005, pp. 748–61.
Lovering, John. 'Postmodernism, Marxism, and Locality Research: The contribution of critical realism to the debate'. *Antipode*, vol. 21, no. 1, 1989, pp. 1–12.
McCalman, Janet. 'Class and respectability in a working class suburb'. *Australian Historical Studies*, vol. 19, no. 74, 1982, pp. 90–103.
McCarty, J.W. 'Australian Regional History'. *Australian Historical Studies*, vol. 18, no. 70, 1978, pp. 88–105.
McCulloch, J. 'The Mine at Wittenoom: Blue asbestos, labour and occupational disease'. *Labor History*, vol. 47, no. 1, 2006, pp. 1–19.
McGerity, L. '"White Queensland": The Queensland government's ideological position on the use of Pacific Island labourers in the sugar sector 1880–1901'. *Australian Journal of Politics and History*, vol. 52, no. 1, 2006, pp. 1–12.
McLeary, M. *Port Pirie Uranium Treatment Plant: Management Plan, Phase 1 – Preliminary Investigation*. Primary Industries and Resources, SA Government, Adelaide, 2004.
McQuade, C.V., Johnston, J. and Innes, S.M. *Review of Historical Literature on the sources and quality of effluent of the Mount Lyell mine site*. Supervising Scientist Report No. 104, Canberra, 1995.
Massey, Doreen. 'The political place of locality studies'. *Environment and Planning A*, vol. 23, 1991, pp. 267–81.
Maynard, E.J., Franks, L.J. and Malcolm, M.S. *The Port Pirie Lead Implementation Program, Future Focus and Directions*. Department of Health, Adelaide, 2005.
Meinig, D.W. 'A Comparative Historical Geography of Two Railnets: Columbia Basin and South Australia'. *Annals of the Association of American Geographers*, vol. 52, no. 4, 1962, pp. 394–413.
Memmott, P. 'From the 'Curry to the 'Weal: Aboriginal town camps and compounds of the western back-blocks'. *Fabrications*, no. 7, August 1996, pp. 1–50.
Menghetti, Dianne. 'Mount Isa: A Town Like Alice?'. *Australian Historical Studies*, vol. 109, October 1997, pp. 21–32.
Moran, M. and Burgen, D. *Gulf and West Queensland Regional Homeland Plan: Stage II Main Report*. Gulf Aboriginal Development Company/ATSIC Gulf and West Queensland Regional Council, Mount Isa, 2000.
Mouat, Jeremy. 'The development of the flotation process: Technological change and the genesis of modern mining'. *Australian Economic History Review*, vol. 36, no. 1, 1996, pp. 3–31.
Mount Isa Welfare Council. *A Report into Some Aspects of Deprivation in an Area of North-West Queensland*. Mount Isa Welfare Council, Mount Isa, 1976.
Mount Lyell Mining Lease Cultural Heritage Assessment Study, Vol. 1: Main Report. Report prepared for Mineral Resources Tasmania and the Department of Environment and Land Management, Godden Mackay Pty Ltd, May 1994.

Bibliography

Murphy, D.J. 'The Dawson Government in Queensland, the first labour government in the world'. *Labour History*, no. 21, May 1971, pp. 1–8.
Murphy, John. 'The Voice of Memory: History, autobiography and oral Memory'. *Australian Historical Studies*, vol. 22, no. 87, 1986, pp. 158–64.
NSW Government Lead Issues Paper. NSW Government Printer, Sydney, 1992.
'Nickel Statistics, US Geological Survey'. 16 December 2008.
O'Farrell, Patrick. 'Oral History: Facts and fiction'. *Quadrant*, November 1979, pp. 2–12.
Onyx, Jenny, Edwards, Melissa and Bullen, Paul. 'The Intersection of Social Capital and Power: An application to rural communities'. *Rural Society*, vol. 17, no. 3, 2007, pp. 215–30.
Palmtree Wutaru Aboriginal Corporation for Land and Culture (in association with Central Queensland Cultural Heritage Management. 'Report on the Aboriginal Cultural Values of the Area to be affected by the Proposed Redevelopment of the No. 7 Dam, Mt Morgan, Queensland'. Prepared for Mcintyre and Associates, Brisbane, 1997.
Passerini, Louisa. 'Attitudes of Oral Narrators to their Memories: Generations, genders, cultures'. *Oral History Association of Australia Journal*, no. 12, 1990, pp. 14–24.
Penrose, B. 'Occupational lead poisoning at Mount Isa mines in the 1930s'. *Labour History*, no. 73, November 1997, pp. 123–41.
Petrow, Stefan. 'Saving Tasmania?: The anti-transportation and Franklin River campaigns'. *Tasmanian Historical Studies*, vol. 14, 2009, pp. 107–36.
Port Pirie Mayoral Reports, 1908–1925. Mortlock Library, South Australia.
Reeves, K., Eklund, E., Reeves, A., Peel, V. and Scates, B. 'Rethinking the international significance of the material culture and intangible heritage of the Australian labour movement'. *International Journal of Heritage Studies*, vol. 17, no. 4, 2011, pp. 301–17.
Richards, Anthony and Robinson, Tim (eds). *Proceedings of the Asset Prices and Monetary Policy – Reserve Bank Annual Conference*. Sydney, 18–19 August 2003.
Richardson, Peter. 'The Origins and Development of the Collins House Group, 1915–1951'. *Australian Economic History Review*, vol. 27, no. 1, 1987, pp. 3–29.
Segal, Naomi. 'Compulsory Arbitration and the Western Australian Gold-Mining Industry: A re-examination of the inception of compulsory arbitration in Western Australia'. *International Review of Social History*, vol. 47, no. 1, 2002, pp. 59–100.
Sorrell, G.H. 'The Dispute at Mount Isa'. *Australian Quarterly*, no. 36, 1965, pp. 22–33.
Sticht, Robert. 'The Development of Pyritic Smelting: An outline of its history'. Presidential Address published in the *Proceedings of the Australian Institute of Mining Engineers*, vol. II, no. II, 1905, Arnell & Jackson/Australian Institute of Mining Engineers, Melbourne, 1905, pp. 9–78.
Teather, E.K. 'Remote Rural Women's Ideologies, Spaces and Networks: Country Women's Association of New South Wales, 1922–1992'. *Journal of Sociology*, vol. 28, no. 3, 1992, pp. 369–90.
Trigger, David. 'Mining, landscape and the culture of development ideology in Australia'. *Ecumene*, vol. 4, no. 2, 1997, pp. 161–80.
Urry, John. 'Localities, regions and social class'. *International Journal of Urban and Regional Research*, vol. 5, 1981, pp. 455–74.
Ville, S. and Merrett, D. 'The Development of Large Scale Enterprise in Australia,

1910–64'. Faculty of Commerce Papers, University of Wollongong, 2000.
Wagner, J. 'Why garden?: Gardening on mining fields in the dry tropics of Queensland, 1860 to 1960'. *Journal of Australian Studies*, vol. 34, no. 3, 2010, pp. 347–61.
Wanna, John. 'A Paradigm of Consent: Explanations of working class moderation in South Australia'. *Labour History*, no. 53, November 1987, pp. 54–72.
Webster, Barbara. 'A "Cosy Relationship" If You Had It: Queensland Labor's arbitration system and union organising strategies in Rockhampton, 1916–57'. *Labour History*, no. 83, November 2002, pp. 89–106.
Webster, Barbara. '"To Fight against the Horrible Evil of Communism": Catholics, community and the Movement in Rockhampton, 1943–57'. *Labour History*, no. 81, November 2001, pp. 155–73.
Weston, Tony. 'Mining Lower Grade Ore: Changes in mining technology at Mount Lyell, Tasmania, 1927–1939'. *Journal of Australasian Mining History*, vol. 8, September 2010, pp. 172–83.
Williams, J. Len. 'The Port Pirie Story: From settlement to city' (updated and amended by Malcolm C.M Bottral). Port Pirie, 2007 [1976].
Wingerd, M.L. 'Rethinking Paternalism: Power and paternalism in a southern mill village'. *Journal of American History*, vol. 83, 1992, pp. 872–902.
Yellowlees P.M. and Kaushik, A.V. 'The Broken Hill Psychopathology Project'. *Australian and New Zealand Journal of Psychiatry*, vol. 26, no. 2, 1992, pp. 197–207.

Unpublished theses

Bertola, Patrick. 'Kalgoorlie, Gold, and the World Economy 1893–1972'. PhD thesis, Curtin University of Technology, 1993.
Cosgrove, B. 'Mount Morgan: Images and realities – dynamics and decline of a mining town'. PhD thesis, CQU, 2001.
Cushing, Nancy. 'Creating the Coalopolis: Perceptions of Newcastle, 1797–1940'. PhD thesis, University of Newcastle, 1996.
Johnstone, C.B. 'A History of the Origins, Formations and Development of the BHAS Ltd at Port Pirie, South Australia'. PhD thesis, University of Melbourne, 1983.
Kirkman, Noreen. 'Mount Isa Mines' Social Infrastructure Programs, 1924–1963'. PhD thesis, James Cook University, 2011.

Acknowledgments

My most important debt is to the people of Broken Hill, Kambalda, Mount Isa, Mount Morgan, Port Pirie, and Queenstown, especially those who agreed to be interviewed. They were patient and welcoming to a tentative outsider with a lot to learn. A large number of museum volunteers, local government staff (especially librarians), as well as local and family historians too numerous to list were always willing to help. I shared the entire journey of researching and writing this book with Antoinette Eklund, whose ideas and input inform this book – and indeed everything I write. Eva and Oliver Eklund were there too. Field trips were often family affairs, and I appreciate their wide-eyed enthusiasm for long journeys and places not on the usual tourist routes.

The University of Newcastle supported this research during its initial stages. During this time Peter Henderson capably and cheerfully assisted with my teaching responsibilities. A Fellowship at the Humanities Research Centre at the Australian National University in 2005 was of great assistance in making a serious start on the research. The entire project was supported by an Australian Research Council's *Discovery Projects* funding grant (project number DP0556725). I was lucky enough to have two excellent research assistants, Shawn Sherlock and Robert Bollard. Betty Cosgrove offered advice on the Mount Morgan chapter, and Paul Adams provided feedback on the Broken Hill chapter. Thanks to the editor of the journal *Labour History*, who gave me permission to reproduce and adapt material I published in that journal in 1999 for Chapter 5. Thanks also to Steven McEachern, from the Australian Social Science Data Archive at the Australian National University, who assisted with access to historical census data. Thanks to oral historians, especially Ed Stokes and Barry York, and to the library staff at Central Queensland University, who facilitated access to the Mount Morgan oral history project interviews.

The latter stages of research and writing were completed while juggling my research and Head of School duties at Monash University. I thank all my colleagues at Monash, especially Louise Bye, Rae Frances, Roberta West, Pam Williams and library staff at the Gippsland and Berwick campuses. David Roberts from the University of New England read and commented on the draft manuscript.

Finally, I'd like to thank the team at NewSouth Publishing especially Phillipa McGuinness for her advocacy and support, Sarah Shrubb for high-quality editorial assistance, Melita Rogowsky for project management and Josephine Pajor-Markus for her design work. An anonymous reviewer from NewSouth Publishing also offered a number of valuable suggestions. Despite these many debts, I am responsible for the final product.

Index

35-hour week campaign 6, 45
360 Collins Street, Melbourne 24

Aaron Hirsch und Sohn 14, 72, 85
Aboriginal peoples, *see* Indigenous peoples
Aborigines Protection Act 1909 (NSW) 50
accommodation, *see also* housing for workers
 boarding houses 42, 77, 152
Addison, Betty 188
Addison, Noel 187, 189
Adelaide 139
Adolphous William mine 20
'Afghan' immigrants 31, 51
air pollution, *see* atmospheric pollution
airfares, payment of for employees 202
alcohol abuse 64, 143, 158, 191
Alma 39, 48–9, 54–5
aluminium, competes with copper 99
Amalgamated Miners' Association 41, 46, 120, 122
Amalgamated Mining Employees' Association 122
Amalgamated Smelters' Association 144
Amalgamated Society of Engineers 145
Amalgamated Workers' Association 144
American Smelting and Refining Company 24, 180, 182
Anglican Church 75, 112, 121, 126–7
Anglo-Celtic immigrants
 at Mount Isa 196
 at Mount Morgan 96
 at Port Pirie 157
 Cornish miners 42, 46
apprenticeships 95, 226
Arbitration Courts 29, 45, 86–7
Argent Street, Broken Hill 38–9
Argus, The 87–8
art galleries 61
asbestos 168, 212
Asian immigrants 51, 82, 157
atmospheric pollution, *see also* lead poisoning; lung conditions

 at Mount Isa 191
 at Mount Morgan 94
 at Port Pirie 142–3, 169
 at Queenstown 132–3
 in mining towns 25
Austin, J.K. 221
Australian Gross Domestic Product 12
Australian Iron and Steel 21
Australian Labor Party
 at Broken Hill 45–7
 at Mount Isa 179
 at Mount Morgan 81, 86
 at Port Pirie 145
 at Queenstown 121–2
 in Tasmania 123
Australian Rules football 49, 125–6, 141, 161
Australian Workers' Union
 at Kambalda 228
 at Mount Isa 179
 at Mount Lyell 123
 at Mount Morgan 86
 at Port Pirie 144, 153
 dominates mining industry 30
Australian–Greek Migration Agreement 160

badge show days 59
Badham, E.T. 'Taffy' 92, 181
Baillieu, M.L. 24
Baillieu, W.L. 16, 230
Bakery and Co-operative Store, Port Pirie 156
Ballarat 12, 20
Baree 97
Barkindji people 49
Baroota 138
Barrier Amalgamated Miners' Association 145
Barrier Daily Truth 52, 60, 62, 64–5
Barrier Highway 37
Barrier Hotel, Port Pirie 157, 164
Barrier Industrial Council
 criticisms of 65
 dominance over Broken Hill 58–60
 emergence of 46
 favours local employment 48
 preference for Anglo-Celtic workers 52
 preferential employment clauses 157

 support for Mount Isa strikers 201
 supports racial integration 51
Barrier Miner 52, 62, 181
Barrier Ranges Miners' Association 42
Barrier Workers' Association 46
Bartley, Frank 40, 42
Battle Mountain massacre 175–6
Bayali people 83
Bayley, Arthur 210
Beattie, George 100
Behind the Barrier 65
Bendigo 12
Benefit Funds and Societies 44, 148, *see also* welfarism
Bennett, Peter 49, 55, 60, 62–4
BHP
 acquires WMC 230–1
 dominates Broken Hill 19
 forms BHAS 23
 foundation of 17, 39–40
 Port Kembla steelworks 6–7
 smelting at Port Pirie 138, 146
 Whyalla steel works 167
'Big Strike' at Broken Hill 29–30, 36, 41–2
Black Friday mill closure 100
Black Star lode 190
Blackwater 100
Blainey, Geoffrey 106
Blanchard, Roland 183
Board, Mary 178
Board of Trade, South Australia 152
boarding houses 42, 77, 152
Bottom, Bob 65–6
Boucher, George 41
Boulder 209–10
Bouldercombe 73
Boulia 195
Bowes Kelly, Anthony Edward 17–18, 20–1, 107, 113
Boyce, James 105
Boyd, Adam 70, 83, 88, 91
Brand, David 215
Brickhill, George R. 143
Bridges, Roy 62
Britain, *see* Anglo-Celtic immigrants; United Kingdom
British Blocks Company 23, 138
Brodie-Hall, L.C. 221

291

Broken Hill 35–67
　35-hour week award 6
　'Big Strike' 29–30
　decline of 32
　early development 12, 17–18
　European immigrants in 31
　fire damage 77
　gender proportions in 54
　geographical characteristics 36–8
　heritage strategy 238
　historical research in 241–2
　lead poisoning treatments 170
　Maltese immigrants 26
　metals mined at 3
　ore body 18
　population levels 35, 37, 48, 54
　Port Pirie linked to 138, 140–1, 145, 165
　support for Mount Isa strikers 201
　workforce declines in 15
Broken Hill Associated Smelters Pty Ltd
　company town approach 234
　dominates Port Pirie 137, 146–7, 156–7
　football team 161–2
　formation of 15, 23
　labour relations under 145, 150, 160–1
　Picnic and Sports Committee 151
　reduces production 168
Broken Hill North Mine 54
Broken Hill Proprietary Limited, see BHP
Broken Hill South Proprietary Limited 19, 23, 38–9, 66
Brookfield, Percy 41, 45, 47
Brooks, Frederick 48
Brown, Dorothy 125–6
Browne, Helen 183–4
Bulali people 49
Bundoona, see Mount Morgan
Bunny, Bruce Frederick 79
Burgen, Dell 194
Burketown 177
Burnie 108–9, 165
Burra 10, 139
Butcher, Bill 89–90

cadmium contamination 169
Callan, Alfred J. 74, 81
Camooweal 176, 194–5
Campion Society 47
Cann, John Henry 46
Cape Horn mine 131
Caputo, Vito 158
Carlton House at Mount Morgan 76, 80, 85, 122

Carpenters' Union 86
Carrow, M.J. 183
Carrow, M.R. 183
Carter Holt Harvey 100
'Casa Grande' 193
Castle, Rob 194
cathode copper 114–15
Catholic Church
　at Broken Hill 44
　at Mount Isa 178
　at Mount Morgan 75, 78
　at Port Pirie 160
　at Queenstown 112, 121, 126
　at Silverton 44
　football league 162
　labour movement and 46–7
Central Queensland coalfields 100
Cester, Lorenzo 56
chain migration 124
Charters Towers 20, 73
children in mining towns 30, 165, 170
Chinese immigrants, see Asian immigrants
churches, see religious denominations
civic buildings
　energy directed to 5
　heritage listing for 66–7
　in Broken Hill 61–2
　libraries 61–2, 127
　Mechanics Institutes 62
　Schools of Arts 78, 81–2, 90
　Town Halls 66
class issues
　at Mount Isa 201
　at Mount Lyell 121
　in Broken Hill 61
　working class culture 28–9
climate
　Broken Hill 36
　Kambalda 207
　Mount Isa 173–4
　Mount Morgan 69
　Port Pirie 135
　Queenstown 105
Cloncurry 176–7, 194, 196
Cloncurry Shire 192
Cobar 20, 37, 164
Cochrane, Peter 15
Cockburn 36, 139
Cockle Creek 170
coke production, at Port Kembla 110
Collins House Group
　BHAS in 23
　dominates mining industry 19
　reformation of 16
　welfarism in 121
　WMC in 210
colonial governments, see State government interventions

Combined Unions Council 198
Commonwealth government, supports Mount Lyell mining 130–1
communism, fear of 201
company picnics 150–1, 155
'company town' approach 146–7, 182–4, 207, 218, 234
Comstock mining 107, 131
Considine, Mick 45–6
contract work 17, 114, 199
Conzinc Rio Tinto Group 16
Cook, Kenneth 62
Coolgardie 207, 209
Coolgardie Shire 231
Co-operative Council 147
Co-operative Housing Societies 192
Co-operative Store, Port Pirie 152–5
Copeman, Charles 102
Copper Bounties Act 1958 (Cth) 93
Copper Mines of Tasmania Pty Ltd 131, 133
copper prices
　1916: 114–15
　1918: 85
　1968: 99–100
　1970s: 129
　postwar rise in 93, 184–5
copper production
　at Kambalda 210
　at Mount Isa 190
　at Mount Lyell 114–15, 131–2
　at Mount Morgan 71–2
　cathode copper 114–15
　in Queensland 175–6
　in South Australia 10, 139
　in Tasmania 107
Corbould, William H. 18, 24
Cornes, T. T. 87
Cornish miners 42, 46
corroborees, money raised by 50
cost of living 45, 190
Country Women's Association 180
Courbold, William H. 24, 181
Cowan, Gladys 164
Cowap, Henri 81
Cowcill, George 210
CRA 16
Crotty (township) 109
Crotty, James 109
Crown Lyell mines 121
Crystal Brook 139, 151, 155
Crystal Brook station 138
CSIRO research into soil contamination 169
cultural issues 27, 233–4, see also ethnicity; European immigrants; welfarism

Index

Daily Truth 52, 60, 62
Dale, George 40
Daniels, Kay 105
Darcy, W.D. 71
D'Arcy, William Knox 19
Darling River 35, 37
Davison, Clem 53
death and injury rates
  at Broken Hill 40
  at Mount Lyell 114
  at Mount Morgan 86, 93–4
  in mining towns 29–30
D'Ebro, Charles 18
Dee River 103
Delamore, William 42–3
Democratic League 81
dental clinic at Port Pirie 153–4
Dial, George 97
diphtheria at Mount Isa 193–4
Djarra 195
Domadgee 195
domestic violence 64
Donley, Robert 168
Duchess copper mine 176, 178
Duchess to Mount Isa Railway Act 1925 (Qld) 176, 185
Dundas 108
Dunne, John 44
dust, prevalence of 25, 30, 36, 94, *see also* atmospheric pollution
Dutton, George 50
dwellings, *see* housing for workers
Dyster, Barrie 13

economic crises 13, 88–91, 236
Edes, Nydia 59
Eight Hour Day celebrations 28
Electricians' Union 86
Electrolytic Refining and Smelting Company 22, 115
Electrolytic Zinc 121
Ellem, Bradon 47
Ellen Street, Port Pirie 142, 157–8, 166
Emerald 100
Empire Hotel, Queenstown 127
employment, *see* labour forces
Emu Bay Railway 108–9, 129–30
environmental pollutants 67, 130, 238, *see also* atmospheric pollution; noise pollution
Esperance 13, 211
ethnicity; *see also* Asian immigrants; European immigrants; Indigenous peoples
  culture and 31
  earnings and 162–3
  in Broken Hill 57
  in Mount Morgan 96
  Queenstown 120
European immigrants 202–3
  from Germany 51
  from Italy 31, 56, 118–19,

158–9
  from Malta 26
  in Broken Hill 51–3, 57
  in Mount Isa 31, 188–9, 195–6, 202–3
  in Port Pirie 151, 157, 160–1, 163
  in Queenstown 116–19, 124
  marginalisation of 159
Evening News 91

family support
  at Broken Hill 61
  at Mount Isa 182
  at Mount Morgan 95
  at Port Pirie 151
  in mining towns 29
Farajoni, Filippo 119
Faull, H.G. 128
Federal government, supports Mount Lyell mining 130–1
Federated Engine Drivers and Firemen's Association 46, 131
Federated Society of Iron Shipbuilders and Boilermakers 145
Federation, economic effects of 12
female labour force, *see also* gender roles
  accommodation industry 77
  at Broken Hill 40, 66
  at Kambalda 219, 222
  at Mount Morgan 86, 92, 95
  at Port Pirie 143
  at Queenstown 117, 124
  employment status 59
  in 1900: 29
feminism 29
Ferguson, John 74–5
Ferguson, Peter 138
fertiliser production 110
Film Broken Hill 62
Finlay, Harry 51–2
Finnish immigrants to Mount Isa 31, 195–6, 202–3
fire damage 77–8, 88–90, 111, 114
Firth, Elizabeth 1, 28, 31–2, 42, 55
Fisher, George R. 181, 193
fishing industry 158
'flawed rise' narrative 102
Fletcher's Creek scheme 96
Flinders Ranges 135
flotation process 16
fly-in-fly-out workers 237, *see also* present-day mining boom
Ford, William 210
Forde, F.M. 98
foreign workers, *see* Anglo-Celtic immigrants; Asian immigrants; European immigrants

forestry in Tasmania 108, 117
Forgan's foundry 138
Fox, Charlie 106, 114, 121
Fraser, Colin 16, 24, 148, 154, 230
fraternal organisations, *see* friendly societies
free airfare offer 202
Freemasonry, among salaried employees 79
friendly societies 44, 46, 60, 78, 112
Frost, Lionel 4, 14
fumes, *see* atmospheric pollution

Galangu peoples 208
Galley museum 132
gambling at Broken Hill 62–3
gas lighting at Mount Morgan 90
Geddes, Charles 145
Geddes, James Sim 150, 165
gender roles 29, 53–4, *see also* family support; female labour force
General Workers' Union 86
geography, economic factors in 16–17
geological exploration 11, 16
German firms in Australia, wartime closure of 23, 85
German immigrants 51
Gladstone 100
Gold Mining Industry Assistance Act 1954 (Cth) 93
gold production
  at Kambalda 205, 208–10, 228
  early discoveries 10
  in Tasmania 106, 108
  in western NSW 37
  migration due to 10
  near Mount Morgan 73
  price trends 93
Golden Eagle nugget 210
Goodhart, J.C. 61
Gordon, John Coutts 73
Gormanston 109, 116, 122, 126
Gray, Robin 131
Great Boulder company 213
Great Depression 89–90, 236
Great Strike at Broken Hill 45, 140
Great War 14–15, 23, 84–5, 108, 114, 121
Greek Orthodox Church in Port Pirie 160
Green, Frank 163, 169
Gregson, Sarah 52
Grey, Fred 145
Gross Domestic Product, contribution of mining 12
'Groupers' 47, *see also* Catholic Church

293

Hagan, Jim 194
Hall, James Wesley 75–6, 79–80
Hall, Thomas Skarrat 19, 71, 75, 79
Hall, Walter Russell 75, 79, 82
Hannan, Patrick 210
Hargreaves, Bryce 94–5, 100
Harvester Judgement 29
Harvey, David 234
health and safety campaigns, see occupational health and safety
Heinkel, Herman 90
Hercus, Louise 138
heritage management
　at Broken Hill 66
　at Mount Isa 194
　at Port Pirie 170–1
　at Queenstown 132
　mining towns 238
Hibernian Catholic Benefit Society 44, 78, 112
historical research 238–9
Horner & Co 142
Hospital Union, Queenstown 113
Housewives Association 202
Housing Commission 197
housing for workers, see also boarding houses
　at Broken Hill 55–7
　at Kambalda 212, 216–18, 224–5, 227
　at Mount Isa 178–9, 182, 192–3, 197, 200–1
　at Mount Morgan 76, 92, 98, 101
　at Port Pirie 144, 152
　at Queenstown 124–5
　in mining towns 235
　tent houses 193–4
H.R. Nicholls Society 229
Hudspeth, Geoffrey 127–8
Hudspeth, L.K. 127–8
Hudspeth, W.L. 127
Hughenden 177
Hughes, William 14
Hunt, Charles 208, 216
Hunt, Graeme 212–13
Huxley, Elspeth 65, 213
Hydro-Electric Commission of Tasmania 130–1
Hytten, Torliev 45

Idriess, Ion 62
Indigenous peoples
　at Broken Hill 49–51
　at Kambalda 208, 222
　at Mount Isa 175, 194–5
　at Mount Morgan 83, 101–2
　at Port Pirie 137–8
　in mining towns 236
　in Tasmania 117–18
　relations with newcomers 5

Industrial Code of South Australia 152
Industrial Commission of Queensland 88, 198
Industrial Conciliation and Arbitration Act (Qld) 198
Industrial Council 46, 157, 201
Industrial Court of Queensland 191
industrial disputes
　at Broken Hill 29–30, 36, 41–2, 45, 140
　at Kambalda 229
　at Mount Isa 179, 198–201
　at Mount Morgan 87
　at Port Pirie 145
　over health and safety 29–30
Industrial Labour Party 41
industrial relations, see Arbitration Courts; industrial disputes; unionism; welfarism
Industrial Relations Commission of WA 228
Industrial Union of Australia 46, 200
industrialisation 7–16, 25–33, see also technological change
injury rates, see death and injury rates
inland corridors 4
International Workers of the World 199
Ireland, immigrants from, see Anglo-Celtic immigrants
'Iron Blow' 106
Italian population, see also European immigrants
　in Broken Hill 56
　in Port Pirie 31, 158–9
　in Queenstown 118–19
itinerant workers 7, 12, 47, 148, 235, see also contract work

Jamestown 139, 142, 168
Jeffrey, F.A. 150
Jenkins, J.F. 155
John Lysaght's 15
John Pirie 137
Johnstone, L.A. 'Lew' 47
Jones, Margaret 27–8, 187–8, 193, 203
Jones, P.V. 228
Jones, Sandy 189, 191, 202

Kabra 74
Kadina 1, 10, 28, 46, 139
Kalgoorlie
　climate of 207
　decline of 213
　growth of 209–10
　Kambalda vs 221
　links to Kambalda 14
　School of Mines 12

Kalkadoon people 175
Kambalda 205–32
　12-hour shifts in 238
　development of 2–3, 24–5
　dust in 25
　gold prospecting at 17
　Kalgoorlie linked to 14
　Kambalda West 216–18
　rail infrastructure 13
　suburban ideals 32
Kane, Frank 183
Kanowna 208
Kapunda 10
Karlson, J.S. 106
Kearns, Roy 66
Kimber, Julie 47
King, Norma 212–13
King River 104, 106, 108, 128, 133
Kirkman, Noreen 177–8
Knox, William 20–1
Kruttschnitt, Julius 182–3, 186, 193
Kruttschnitt, Marie 184
Kurialda 178
Kwinana refinery 211

Labor Party, see Australian Labor Party
Labour Council 144–5, 201
Labour Day 86, 155–6
labour forces, see also female labour force; staff (salaried employees)
　at Broken Hill 40, 66
　at Kambalda 218
　at Mount Isa 185–6, 191
　at Mount Lyell 115–17, 124, 131–2
　at Mount Morgan 72–3, 76, 84, 86, 90, 92
　at Port Pirie 148, 162
　declines in 15
　development of 11
　Hydro-Electric Commission 130
　in mining, peak levels 14
　preferential employment clauses 58–9
Lake Burbury 110
Lake Lefroy 208
Lake Macquarie, smelting operations 22
Lake Margaret power scheme 130
Lake Moondarra scheme 197
Lambton coal mine 20
Lampen, Sigge 28, 187, 202–3, 239
Largs Bay holiday camp 56
Larkin, Percy 208
Laud, Peter 212
Launceston 106–8
lead poisoning
　at Broken Hill 41

Index

at Mount Isa 177, 179, 191
at Port Pirie 151, 153-4, 169
through workplace exposure 30
lead production
at Mount Isa 183, 186-7
at Port Pirie 161
prices and 54-5, 184-5
wage bonus for 31, 55, 162
Lefroy, Henry Maxwell 208
Leichhardt River 173-4, 178
Leigh, Jim 95
Lewis, Essington 164-5
'life history' approach 240
Linda Valley 106, 109, 116, 118
Lindesay Clark, Gordon 229-30
line of lode, see ore bodies
Lisle, Roger 79
Lithgow 20, 26
'Little Ireland' 78
London Metal Exchange 85, 114, 184
Lower Works, Mount Morgan 71-2
loyalty of employees 80, see also welfarism
Lundager, J.H. 86, 98
lung conditions 30, 40-1, 94
Lunnon, Jack 212
Lyell Highway 108, 128
Lynchford 106, 108, 117

Mackie, Pat 61, 190, 198-200, 202
Macquarie Harbour 109, 118, 133
Mail, The 166-7
Malayan immigrants, see Asian immigrants
Maltese immigrants 26, 51
Manchester Unity Independent Order of Oddfellows 44
Manos, Will 2, 214, 222-3, 226
maps
Australia *frontispiece*
New South Wales 34
Queensland 68, 172
South Australia 136
Tasmania 103
Western Australia 206
maritime labour at Port Pirie 157
Mary Kathleen uranium mine 217
Matilda Highway 186
Maxwell-Stewart, Hamish 105
McAskill, A.F. 181
McCarty, J.W. 4
McCoy, Kelvin 131
McCulloch, George 17-18
McCulloch, James 18
McDonald, Donny 95-6
McDonald, James 121
McDonald, John 121
McDonald, Thomas 121-2

McDonough, Michael 106
McDonough, William 106
McGuire, Paul 36, 57-8
McKinley Highway 197
Mechanics Institutes 62
mechanisation, see industrialisation; technological change
Menindee Mission Station 50-1
Mercury 111, 118-19
Meredith, David 13
Metal Manufactures Pty Ltd 85
metallurgy, technological changes in 15
Methodist Church 46, 51, 75, 178
Michaels, Carole 56, 64-5
migrant workers, see Anglo-Celtic immigrants; Asian immigrants; European immigrants
Miles, John Campbell 174, 176, 210
Miller, Lawrence 200
Milthorpe, George 178-9
Minchin, Eric 61
Mine Managers' Association 30, 44, 58-60
Minerals Mining and Metallurgy Limited 66
miners' lung, see lung conditions
Mineside, Mount Isa 182, 192, 207
mining, development in Australia 7-16
Mining Trust Co 182
Mitke, Charles 183
modernism 188-9, 231, 237
Molfetta, Italy 158
Monckton, William 75, 78
Moonta 10, 139
Moorakyne 18
Moran, Mark 194
Morgan, Hugh 228-9
Morgan, John 210
Morgan brothers 19, 71, 83
Morning Bulletin 76, 80-1, 89
Moroney, John 112
Morris, Albert 56
Moulds, Lorna 94
Mount Bischoff 108-9
Mount Isa 173-204
early development 24
Finnish immigrants 31
geographical characteristics 175-7
growth of 3
industrial disputes 30
itinerant workers 26-7
lockout at 61
ore body at 11
recollections of 27-8
role of State government in 234

Mount Isa Chamber of Commerce 201
Mount Isa Mail 196
Mount Isa Mines Limited
begins 12-hour shifts 203
company town approach 234
formation of 22, 24, 175
housing supplied by 197
profits made by 185
welfarism 179-80
working conditions 189-94
Mount Isa News 196-7
Mount Lyell 105-34, see also Queenstown
Mount Lyell Mining and Railway Company Limited
formation of 20-2
goes into liquidation 131
growth of 109-10
pressures Italians to leave 119
Queenstown dominated by 120, 133-4
working conditions 113-16
Mount Lyell Mining and Railway Company Limited (Continuation of Operation) Acts (Tas) 130
Mount Morgan 69-103
decline of 32
early development 12, 19-20
geography of 73-4
German firms ousted from 14
historical research in 241-2
lights go out in 90
lockout at 30
metals mined at 3
Rockhampton linked to 14
Somerset family in 164-5
South Calliungal 75
Mount Morgan Gold Mining Company 70-1, 88-9
Mount Morgan Limited 22, 70-1, 91, 93, 99
Mount Morgan (West) Gold Mine 70-1
Mount Owen 109, 132-3
mulga, at Broken Hill 39
Mundic Works 72
municipal buildings, see civic buildings
Municipal Employees Union 47
municipal government 27, 38, 137
Murchison Highway 108
Murray, Hugh 127
Murray, R.M. 120, 122, 127
Mussen, Gerald 146-7, 150, 154

Napperby 139, 144
National Trust 132, 170-1, 194, see also heritage management
New South Wales, see also Broken Hill
development in 9-11

295

map 34
rail freight rates 13
Newcastle
 growth of 10
 industrial disputes 61
 links to Port Kembla 7
 Lysaght's moves to 15
 steelworks constructed on waste ground 174
Newtown, R. 84
Newtown, Walter 50
nickel production 210–11, 213, 219–20
Nickel Refinery Act 1968 (WA) 214
Nickel Refinery (Western Mining Corporation) Limited Agreement Act 1968 (WA) 215
Nicklin, Frank 189, 201
night cart system 57
Nisbet, Hume 84
noise pollution 94, 191
Norman, Mary 215
Normanton 176
Norseman 13, 211
North Broken Hill 47–9
North Broken Hill Limited 19, 23, 99
North Lyell Mining Company 107, 109, 114, 121, 124
North Lyell tunnel 116, 131
North Queensland Rugby League Association 180, 195
Northern Argus 70–1
Nukunu peoples 137–8
Nunan, Joy 166
Nurmi, Kalle 189, 195–6

occupational health and safety 29–30, 45, 185, *see also* lead poisoning; lung conditions; state government interventions; welfarism
Oeser, O.A. 220–1
Oliver, J.B. 216, 221
Olsen, Vera 177–8
Olympic Flame club 161
O'Neil, E.P. (Paddy) 46–7
O'Neil, W.S. 'Bill' 47
O'Neil, W.S. 'Shorty' 47, 58, 61, 65
open stope mining 41
oral history interviews 27–8, 38, 65, 240–2
ore bodies 39–40, 173, 190
O'Regan, Michael 112
O'Regan, Patrick 96, 98
O'Reilly, John Joseph 45
Organisation for Tasmanian Development 131
Orr, William 20, 107
'Outback at Isa' exhibition 203
Outokumpo Oy 211

Pacific Island labourers 82
Parbo, Arvi 229
Parker Hill 208
Parkes, Roy 138
Parkside, Mount Isa 192–3
Pastoralists' Association 38
paternalism, *see* welfarism
Pattison, William 19, 71
Peko-Wallsend Ltd 99
'Penghana' (house) 111, 122, 127
Penghana (locality) 3, 111
Pentecostal Church 31, 195
Peterborough (formerly Petersburgh) 36, 136, 139
phthisis, *see* lung conditions
Pickup, John 61
Pieman River 108
Pirie West 144, 161
police intervention in Mount Isa 199–200
political issues 58–60, 79–82, 166–71, *see also* state government interventions; unionism
pollutants 67, 130, 238, *see also* atmospheric pollution; noise pollution
population levels
 Adelaide 139
 Broken Hill 31, 35, 37, 48, 54
 Kambalda 209, 214, 225, 230
 Mount Isa 177, 196
 Mount Morgan 76, 90, 92, 100
 Port Pirie 31, 137–8, 143, 157, 160, 168
 Queenstown 112–13, 124, 126, 129
 Whyalla 161, 167
Port Adelaide 140, 145, 170
Port Augusta 135, 138, 211
Port Flinders 149
Port Germein 140, 144
Port Kembla
 BHP steelworks at 6–7
 coke production 110
 copper smelter 72
 ERS founded at 22
 German firms ousted from 14
 links to Wollongong 21
 loses smelting work to Mount Lyell 115
 Metal Manufactures Pty Ltd founded at 85
Port Pirie 135–71
 early development 3, 14, 22–4
 Italian population 31
 lead poisoning in 30
Port Pirie Advertiser 153
Port Pirie Chamber of Commerce 145
Port Pirie Hotel 142

Port Pirie Lead Implementation Program 170
Port Pirie Railway Act 1873 (SA) 139
Port Pirie River 135
Port Pirie Trades and Labour Council 145
Port Pirie Wharf Company 146
Poseidon float 220
preferential employment clauses 58–9, 157
Presbyterian Church 75, 112, 126
present-day mining boom 6, 202, 237
Primitive Methodism 46, 75
Prince Lyell mine 107, 131
Princess River 133
Progress Associations/Committees
 at Broken Hill 42–3, 48, 67
 at Kambalda 209
 at Mount Isa 178, 180
 at Queenstown 111
 at Silverton 42
prostitution 64–5, 157–8
Pryor, Jenny 170
Purnamoota 42–3
pyritic furnaces 26, 110

Queen Elizabeth II 193
Queen River 25–6, 105–6, 108, 111, 133
Queensland, *see also* Mount Isa; Mount Morgan
 map 68, 172
Queenstown 105–34
 decline of 32
 early development 12, 14, 20–2
 fire damage 77
 metals mined at 3
 state government interventions 234
 water supply 25–6
Queenstown Football Association 125–6
Quinn, Albert 118–19

racial discrimination 52–3, 82, 158–9, *see also* ethnicity
Radium Hill 167–8
rail infrastructure
 at Broken Hill 43
 at Kambalda 215
 at Mount Isa 177, 190
 at Mount Morgan 74, 85, 98
 at Port Pirie 22–3, 139, 141–2, 162, 168
 development of 10
 in New South Wales 13
 in Tasmania 107–10, 129–30, 132
 in Western Australia 211

Index

Rasp, Charles 17, 37, 210
rats in mines 94
Recorder, The 143
Red Hill, Mount Morgan 76, 78, 97, 208
Red Hill (WA) Gold Syndicate 209
Redding, P.J.R. 221–2
Redman, Janice 125, 129
Reece, Eric 130
regeneration of mining areas 67, 132–3
regional capital 17, 20
religious denominations, *see also names of denominations*
  at Broken Hill 43
  at Mount Isa 178, 180, 195
  at Mount Morgan 75
  at Port Pirie 160
  at Queenstown 112
  labour movement and 46
Renison Goldfields Consolidated 129
rent-purchase housing scheme 125
'respectability' 28–9, 149–52
Revenge mine 228
Reynolds, Henry 105
Richards, G.A. 98
Rigg, Gilbert 146
Rio Grande mine 190
Rio Tinto Group 16
Rise of Port Pirie, The 168
Robert Sticht Memorial Library 127–8
Robinette, William 153–4
Robinson, William S. 14, 24, 146, 150–1
Rockhampton 14, 19–20, 74, 100
Rosebery 108
Ross, R.S. 62
Rowley, W. 209
Royal Commission into Lead Poisoning 30, 151, 154, 159, 169
Royal Flying Doctor Service 196
Royal Tharsis mine 124, 131
rugby league 97, 180, 195
Russo-Asiatic Consolidated Ltd 179
Ryan, John Joseph 112

safety campaigns, *see* occupational health and safety
Said, Paul 26
salaried employees, *see* staff (salaried employees)
Salvation Army 75, 126
San Francisco Call 184
Saslavski, Melba 178–9
Scarman, Brian 161–2
Schlapp, Hermann 20
School of Mines 12

Selwyn Ranges 173
Sheil, Glenister 92
Shields, John 31, 47
silicosis, *see* lung conditions
Silver Lake shaft 215
silver-lead mining 54, 108, 185–6, *see also* lead production
Silverton 37–8, 42, 51
Simons, Nicolas 145
slag 165
smelting operations
  at Broken Hill 22
  at Mount Isa 115, 183, 186–7, 199
  at Mount Lyell 129
  at Mount Morgan 99
  at Port Kembla 72, 115
  at Port Pirie 138, 146, 149
  at Whyalla 166–7
  employment in 15, 162
  in Tasmania 110
  in Western Australia 211
Smith, Harry 178
soccer, in Port Pirie 161
soil contamination 169, *see also* environmental pollutants
Soldiers' Hill 192
Soldiers' Memorial Park and Playground, Port Pirie 149–50, 154–5
Solomontown 137, 166
Somerset, Harry 165
Somerset, Henry 164–5
Somerset family 164–5
South Australia, *see also* Port Pirie
  map 136
  migration to Barrier Ranges from 42
  mining in 10
South Australian Railways 43, 162, 168
South Broken Hill 39, 49
South Broken Hill Mine 54
South Calliungal, *see* Mount Morgan
South Queenstown State School 118
spatial fix 234–5
Spencer Gulf 135, 137, 149
sporting activities
  at Broken Hill 48–9
  at Kambalda 227
  at Mount Isa 180, 195
  at Mount Morgan 86, 97
  at Port Pirie 161–2
  at Queenstown 125–6
  Australian Rules football 49, 125–6, 141, 161
  Rugby League 97, 180, 195
  soccer 161
staff (salaried employees)
  at Kambalda 224–5, 227
  at Mount Isa 183

  at Mount Lyell 115–16
  at Mount Morgan 79
  at Port Pirie 163–4
  facilities for 91–2, 127
stamping batteries 71, 107
Stanley, Lord 37
State government interventions
  early support for mining 12–13
  in Broken Hill 60
  in Kambalda 215
  in Mount Isa 177, 185, 199–200
  in Mount Lyell 129–31
  in Mount Morgan 87, 98
  in Queensland 13
  role in industrial development 234
  wages determined by 152
State Transport Act (Qld) 199–200
Steel, Bruce 159–60, 163
stereotyping 202–4
Sterlite Industries (India) Limited 131
Sticht, Marion 110
Sticht, Robert 110–12, 120, 122
Stock, Trevor 101
Stokes, David 218–19, 223–6
Stopford, Jimmy 87, 97–8
Strahan 14, 107–11, 121
Strangways Act (SA) 139
Struck Oil 97
Sturt, Charles 37
suburban ideals 32, 48, 192–4, *see also* family support; housing for workers; welfarism
Sulphide Corporation 22
sulphide ores 211
sulphur dioxide 94, 132–3, 142–3, 169, *see also* atmospheric pollution
Summerton, Harold 155
Sunset, Mount Isa 197
Sydney Morning Herald 64
'Syrian' immigrants 51

Tasmania, *see also* Queenstown
  development in 9
  map 103
  west coast of 107–11
Tasmania Grant (The Mount Lyell Mining and Railway Company Limited) (Cth) 130
technological change, *see also* industrialisation
  at Broken Hill 54–5
  at Mount Isa 189
  at Mount Lyell 124
  at Port Pirie 148, 161
  causes workforce decline 15
  in mining 11
Telowie 160
tent houses 193–4

297

Terowie 139–40
Thackaringa 37
Thalpiri, see Port Pirie
The Argus 87–8
The Mail 166–7
The Recorder 143
The Rise of Port Pirie 168
Thomas, Eric 132
Thomas, Josiah 80
Thompson, Captain 137
Thompson, Max 26–7
Tillet, Ben 123
Timms, Arthur 89, 97
tin production 10, 108–9
Tindale, Norman 138
Tipperary Point 76, 78, 97
title disputes 71
Torres Strait Islander peoples, *see* Indigenous peoples
tourism 97–8, 132
Townside, Mount Isa 192
Townsley, W.A. 122–3
Townsville 185, 190
Townsville Daily Bulletin 180
trams in Broken Hill 43
tree preservation policies 218
Trial Harbour 109
Tullah 108
two-up 63
typhoid outbreaks in Broken Hill 53

underground mining, masculinised nature of 29
UNESCO conference 220–1
unionism, *see also* industrial disputes; welfarism
  at Broken Hill 22, 44, 60–1
  at Mount Isa 181, 198, 201–2
  at Mount Lyell 122–3
  at Mount Morgan 80–1, 92
  at Port Pirie 141, 144, 153
  at Queenstown 120, 122
  health and safety campaigns 41–2
  'respectability' and 28–9
  treatment of 'scabs' 48
United Kingdom 11, 14–15, 30–1, *see also* Anglo-Celtic immigrants
Upfield, Arthur 62
Upper Works, Mount Morgan 72
uranium production 167–8
Urquhart, Leslie 179

Van Diemen's Land, *see* Tasmania
Vaughan, Crawford 145
Vedanta Ltd 131
Verschuer, Jean 216
Virtus Club 161

wage labour 17, 40, 79
wage rates
  at Broken Hill 31, 55
  at Kambalda 228
  at Mount Isa 183, 190–1, 194, 198
  at Mount Lyell 114–15
  at Mount Morgan 94
  at Port Pirie 152–3, 162–3
Wall, Sydney 93–4, 241
Wallaroo 10, 139
Walter and Eliza Hall Trust 82
Walter Hall (locality) 97
Waratah 108
water supplies
  Broken Hill 43, 45, 53
  Kalgoorlie 211
  Mount Isa 183, 197
  Mount Morgan 71, 96
  Port Pirie 142
waterfront at Port Pirie 166
Waterside Workers' Federation 144, 153, 155–6
Watson, May 38, 58
Waxman, Percy 111
Weber, Max 231
Weeroona Holiday Camp 56, 149–50, 154
welfarism, *see also* 'company town' approach
  at Mount Morgan 91
  at Port Pirie 146–8
  at Queenstown 120–1, 123, 126–8
  clinics and hospitals 153–4
  company picnics 91, 150–1, 155
  Hospital and Medical Unions 113, 120
  preferential employment clauses 58–9, 157
  Provident Funds 115, 147
  rent-purchase scheme 125
  Sickness and Accident Funds 147, 153–4
West Australian 212–13
West Coast Premiership 126
West Coast Trades and Labour Council 123
West Lyell open cut mine 124, 131
West Lyell workshop 129
West Works, Mount Morgan 72
Western Australia, *see also* Kambalda
  gold discoveries in 10
  map 206
  rail infrastructure 13
Western Mining Corporation
  company town approach 234
  economic downturn 228
  formation of 22, 24
  Kambalda dominated by 210, 214–16
  research sponsored by 220–2
Weston, Tony 116
wheat cultivation 137
White, Richard 101
White Cliffs 38, 52
White Feather 208
'White Queensland' policy 82
Whitley reports 147
Whyalla 135, 160, 166–7
Widgiemooltha 210
Wightman, T. 81
Wilcannia 50–1
Wiljakali people 49
Williams, Ted 66
Windarra 220
wind-blow material 169
WMC–Lindesay Clark Trust Fund 229
Wollongong 6–7
Woman's Day 190
women, *see* family support; female labour force; gender roles
Women's Weekly 219–20
Wongatha people 222
Woodward, W.A. 154–5
workers' compensation schemes 41, 45
Workers' Education Association 127
Workers' Industrial Union of Australia 46, 200
workforce, *see* labour forces; staff (salaried employees)
working conditions 121, 147–8
Workmens Compensation (Broken Hill) Act 1921 (NSW) 41
Workmens Compensation (Lead Poisoning) Act 1922 (NSW) 41
workplace committees 147
workplace safety, *see* occupational health and safety
Works Council movement 147
works picnics 91
World War I (Great War) 14–15, 23, 84–5, 108, 114, 121
Worthington, Norma 223–4, 226

Yallambee 194
Yallourn 212
Yarraville 110
Yorke Peninsula 10

Zeehan 14, 108–9
Zinc Corporation 15–16, 19, 23, 55–6
zinc production 54, 148, 167
Zubrinich, Leonard 61, 140, 143, 165

www.ingramcontent.com/pod-product-compliance
Lightning Source LLC
Chambersburg PA
CBHW021346300426
44114CB00012B/1104